Muffles' Truffles

Multiplication and Division with the Array

Antonia Cameron
Catherine Twomey Fosnot

DEDICATED TO TEACHERS™

*first*hand
An imprint of Heinemann
361 Hanover Street
Portsmouth, NH 03801–3912
firsthand.heinemann.com

Offices and agents throughout the world

ISBN 13: 978-0-325-01019-9
ISBN 10: 0-325-01019-6

Harcourt School Publishers
6277 Sea Harbor Drive
Orlando, FL 32887–6777
www.harcourtschool.com

ISBN 13: 978-0-15-360571-0
ISBN 10: 0-15-360571-5

The development of a portion of the material described within was supported in part by the National Science Foundation under Grant No. 9911841. Any opinions, findings, and conclusions or recommendations expressed in these materials are those of the authors and do not necessarily reflect the views of the National Science Foundation.

Library of Congress Cataloging-in-Publication Data
CIP data is on file with the Library of Congress

Printed in the United States of America on acid-free paper

14 ML 4 5 6

Acknowledgements

Photography

Herbert Seignoret
Mathematics in the City, City College of New York

Illustrator

Joy Allen

Schools featured in photographs

The Muscota New School/PS 314 (an empowerment school in Region 10), New York, NY
Independence School/PS 234 (Region 9), New York, NY
Fort River Elementary School, Amherst, MA

Contents

Crayola

Unit Overview

The focus of this unit is the development of the open array as a model for multiplication and division. This unit uses a series of investigations based on the context of Muffles' Truffles shop. The questions posed in the first investigation (how many boxes of ten can be made with a given quantity of truffles; how many leftovers will there be from a given quantity and how can they be combined to make assortment boxes; and what is the cost of a given quantity of truffles if they cost $1 each?) give students an opportunity to explore place value—the multiplicative structure of our base-ten system and quotative division. In the second and third investigations, students build two-dimensional blueprints of one-layer boxes and use these arrays to explore some of the big ideas in multiplication (the distributive, associative, and commutative properties). In the fourth and final investigation, students work with open arrays in the context of labeling and pricing wrapped boxes of truffles. To figure out the dimensions of the wrapped boxes (or open arrays) and the cost, students need to apply a number of big ideas previously developed in this unit.

There are three different kinds of minilessons for multiplication included in the unit as well: counting around the circle, strings of related problems,

The Landscape of Learning

BIG IDEAS

- Unitizing
- The distributive property of multiplication (over addition and over subtraction)
- The commutative property of multiplication
- The place value patterns that occur when multiplying by the base
- The associative property of multiplication

STRATEGIES

- Using repeated addition
- Skip-counting
- Using partial products
- Using ten-times
- Doubling and halving

MODEL

- Open array

and quick images. The count-around is used to support the development of place value as it relates to multiplication. The strings of related problems are explicitly designed to guide learners toward computational fluency with whole-number multiplication and to build automaticity with multiplication facts by focusing on relationships. The quick images use 2 × 5 and 1 × 5 arrays as units to build larger arrays. In the last days of the unit, more complex minilessons (double-digit multiplication problems) generate a wider range of student strategies that can be explored (and modeled) with the open array.

BIG IDEAS

This unit is designed to encourage the development of some of the big ideas related to multiplication:

- ***unitizing***
- ***the distributive property of multiplication (over addition and over subtraction)***
- ***the commutative property of multiplication***
- ***the place value patterns that occur when multiplying by the base***
- ***the associative property of multiplication***

The Mathematical Landscape

Research has documented that different mathematical models have different effects on mental computation strategies (Beishuizen 1993; Gravemeijer 1991). For addition and subtraction, the open number line is well aligned with students' invented strategies: it stimulates a mental representation of numbers and number operations that is powerful for developing mental arithmetic strategies because students using the open number line are cognitively involved in their actions. In contrast, students who use base-ten blocks or the hundred chart tend to depend primarily on visualization, which can result in a passive "reading off" behavior rather than cognitive involvement in the actions undertaken (Klein, Beishuizen, and Treffers 2002).

For multiplication and division, the open array can play a similar role. Partial products can be represented on it, and the relationship of these to the total area can be explored to develop an understanding of the distributive and associative properties. Research by Battista et al. (1998), however, suggests that this model is often difficult for learners to understand because the ability to coordinate rows and columns simultaneously requires a substantial cognitive reorganization—and thus an understanding of arrays itself goes through successive stages of development. This unit is designed to support learners in developing a deep understanding of the array, so that eventually they will be able to use it as a powerful tool to think with for multiplication and division.

Unitizing

Initially, students think about multiplication additively. They use repeated addition or skip-counting. To really think multiplicatively, they have to learn to think about the group as a unit. This process of making the group a unit is called unitizing. When students are first introduced to multiplication, most have just reached solid ground with counting. They do know what six is; it is a group of six objects. And they know what three is; it is a group of three objects. But to think of 3 × 6, they have to cognitively reorganize everything they know about number. They have to think of the group of six as one unit because they need to make three groups of six. To a young learner, this makes no sense. In fact, it seems to be a contradiction—something can't be six and one at the same time!

The distributive property of multiplication (over addition and *over subtraction)*

The distributive property of multiplication over addition—knowing that factors can be broken up and distributed to make partial products, which can then be added together to produce the product of the original factors—is one of the most important ideas about multiplication. It is the reason that the standard pencil-and-paper algorithms for multiplication and division work, and its use for mental arithmetic is powerful as well.

Algebraically, the distributive property is represented as $a(b + c) = ab + ac$. For example, 4 × 19 can be solved by partitioning the 19 into 10 and 9. Two partial products can now be made from 4×10 and 4×9. Adding those products, $40 + 36$, produces the answer to 4×19. The 19 can also be distributed in other ways, such as 12 and 7, or the 4 can be distributed into two groups to make $19 \times 2 + 19 \times 2$. In these examples, factors were decomposed into addends and distributed, therefore the partial products had to be added, making use of the distributive property of multiplication over addition. Nineteen can also be expressed as $20 - 1$. Now, the partial products can be found by calculating 4×20 and subtracting from it the product of 4×1. This produces $80 - 4$, which also is equivalent to 4×19. In this case, we have used the distributive property of multiplication over subtraction.

For students, an understanding of this property of multiplication can be quite difficult to construct. Understanding that 6×4 means six *groups* of four—not 6 *plus* 4—can itself be elusive until students construct unitizing. Then, with the distributive property, the challenge is even greater. Students have to think of making groups out of the groups! Students need much time to explore, arranging and rearranging the parts of the numbers involved, in order to develop this big idea. This unit gives students many opportunities to explore these ideas.

❖ The commutative property of multiplication

Multiplication is also commutative: $a \times b = b \times a$. Picture 4 towers, each made with 19 connecting cubes. Now imagine them right next to each other so they make a 19×4 array. If we turn this array 90 degrees, we have a 4×19 array—19 towers with 4 cubes in each. Using arrays like this is exactly what students do to convince themselves of the commutative property. They often call it the turn-around rule. There are many opportunities in this unit for students to construct this big idea as they explore arrays and think about how they might prove that a 6×2 blueprint is the same as a 2×6. This idea can be modeled with the array; no matter how the array is rotated, the area resulting from multiplying the dimensions of 2 and 6 is the same.

❖ The place value patterns that occur when multiplying by the base

Precisely because multiplication is commutative, an interesting thing happens when students multiply by the base—the factor "moves over" to the appropriate column. For example, $10 \times 4 = 4 + 4 + 4 + 4 + 4 + 4 + 4 + 4 + 4 + 4$. The result of 40 seems amazing to students, who often say that they added a zero or refer to the zero trick. The reason this "trick" works is that the ten groups of four can also be thought of as four groups of ten—so the four is placed into the tens column. It is important to support students in exploring why this pattern occurs—to help them construct how place value is involved.

❖ The associative property of multiplication

The associative property also holds for multiplication: $(a \times b) \times c = a \times (b \times c)$. Picture a 4×5 array composed of two 2×5 arrays. The two 2×5 arrays can also be rearranged into a 2×10 array. In a sense, this is a specific case of the associative property: $(2 \times 2) \times 5 = 2 \times (2 \times 5)$. Although the generalization of the associative property is not the goal of this unit, specific cases like this are likely to come up. The associative property is revisited in the unit, *The Box Factory*—a nice unit to use immediately following this one.

STRATEGIES

As you work with the activities in this unit, you will notice that students use many strategies to solve the problems posed to them, including:

- ***using repeated addition***
- ***skip-counting***
- ***using partial products***
- ***using ten-times***
- ***doubling and halving***

❖ Using repeated addition

Often the first strategy students use to solve multiplication problems is repeated addition. This is because they are viewing the situation additively, rather than multiplicatively. To solve 6×4 (for example, how many truffles are in a one-layer box), students will write (or think about) $4 + 4 + 4 + 4 + 4 + 4$, and then add to find the total. Repeated addition should be seen as a starting place in the journey, but not as the end point. As you confer with students, you will need to help them keep track of the groups, and you can encourage more efficient grouping (such as turning the six fours into three pairs of fours, resulting in three eights).

❖ *Skip-counting*

The struggle to keep track of the groups usually pushes students to skip-count. Although they are still thinking about the situation additively rather than multiplicatively, they keep track of the groups mentally and skip-count. To figure out how many truffles are in a 6 × 4 one-layer box, students might count 4, 8, 12, 16, 20, 24.

❖ *Using partial products*

An important shift in multiplicative thinking occurs when students begin to make partial products—they use a fact they know to make another. For example, they might reason that if a 5 × 4 box holds 20 truffles, and a 2 × 4 box holds 8 truffles, then a 7 × 4 box must hold 28 truffles. Here they are finally making the group a unit (unitizing) and adding it or subtracting it as needed. The big ideas underlying this strategy are unitizing and the distributive property.

❖ *Using ten-times*

Once students begin to make use of partial products, an important strategy to encourage is the use of the ten-times partial product. This can be very helpful for the nine-times table (just subtract one group from ten-times), for the five-times table (half of the product of ten-times), and for double-digit multiplication. Of course, this strategy is helpful only if students have constructed an understanding of the place value patterns that occur when multiplying by the base.

❖ *Doubling and halving*

As students' multiplicative reasoning becomes stronger, they develop the ability to group more efficiently. They begin to realize that if they double the number of groups and want the total product to be the same, they need to halve the amount in each group: 4 × 6 = 8 × 3. This strategy can be generalized to tripling and thirding or quadrupling and quartering, etc. The big idea underlying the reason that these strategies work is the associative property of multiplication. In multiplication, the factors can be associated in a variety of ways. For example, 4 × (2 × 3) = (4 × 2) × 3. These strategies may arise as students build blueprints and think about the dimensions of their open arrays ("wrapped boxes"), but the associative property is not a primary focus of this unit. The unit in the *Contexts for Learning Mathematics* series that focuses on development of the associative property is *The Box Factory*.

MATHEMATICAL MODELING

The primary focus of this unit is the development of the array model for multiplication. An array is a rectangular arrangement of rows and columns. Initially, the array is introduced as the arrangement of truffles in a box, but as the unit progresses, two-dimensional graph paper blueprints of the boxes are used. Eventually, box outlines become the context and the open array replaces the graph paper arrays, making it a model that can be used to examine product and partial product relationships.

The open array is a powerful model for multiplicative thinking because it can support the development of the following:

- a wide range of strategies (skip-counting, repeated addition, doubling, doubling and halving, partial products) and big ideas like the distributive, associative, and commutative properties
- visual representations for what multiplication means (e.g., 6 × 8 can be understood as 6 rows of 8 squares or 8 columns of 6 squares)
- an understanding of area and perimeter

Models go through three stages of development (Gravemeijer 1999; Fosnot and Dolk 2002):

❖ ***model of the situation***

❖ ***model of students' strategies***

❖ ***model as a tool for thinking***

❖ *Model of the situation*

Initially, models grow out of visually representing the situation. For example, in this unit arrays are introduced as an arrangement (rows and columns) of truffles in a box. As students explore and design boxes for the truffles, the blueprint (graph paper arrays) emerges as a representation of the box—and now it is a two-dimensional rectangular array. Quick images are introduced using the 2×5 and the 1×5 blueprints as units, and these arrays depicting the situation become the scaffold that allows students to envision how larger arrays can be made with smaller arrays. Eventually, outlines of the boxes are used in which there are no rows and columns to count.

❖ *Model of students' strategies*

Students benefit from seeing the teacher model their strategies. Once a model has been introduced as a representation of the situation, you can use it to display student strategies. If a student says about solving 12×9, "I used 12×10 and subtracted 12," draw the following:

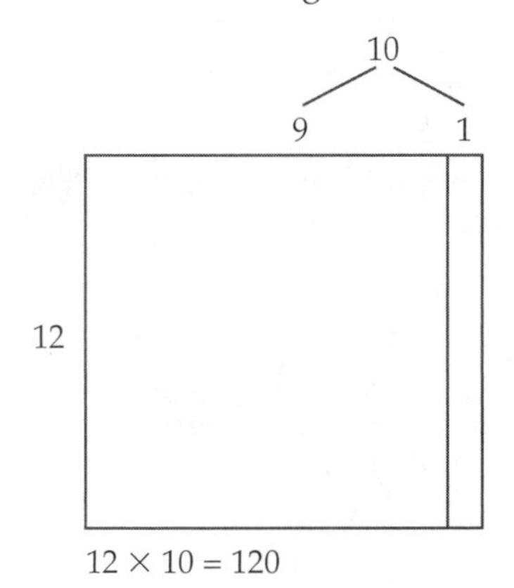

$12 \times 10 = 120$

$12 \times 9 = (12 \times 10) - (12 \times 1) = 120 - 12 = 108$

Another student in calculating 12×9 might say, "I knew 10×9 and 2×9 and I just added them together to get 108." Here you can represent the partial products as follows:

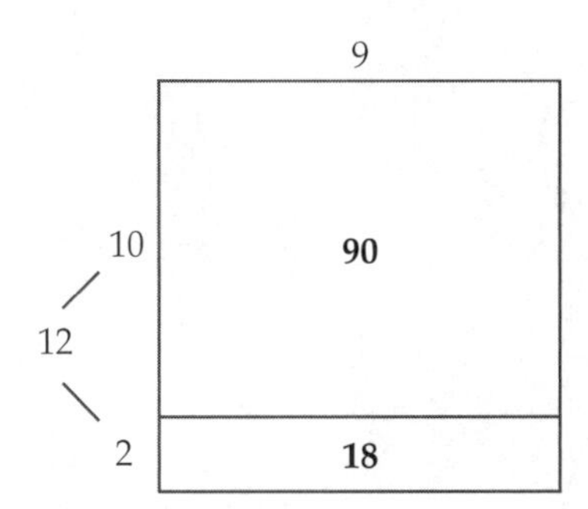

$12 \times 9 = (10 \times 9) + (2 \times 9) = 90 + 18 = 108$

Representations like these give students a chance to see and discuss each other's strategies.

❖ *Model as a tool for thinking*

Eventually, students become able to use the model as a tool to think with—to explore and prove their ideas about multiplicative reasoning. Here, as the dimensions of a box are visualized in relationship to the dimensions of other boxes, students can explore the distributive property of multiplication over addition (or subtraction), the associative property, and the commutative property.

1. The distributive property of multiplication over addition: the original 2×5 open array is used to figure out the dimensions of a larger open array:

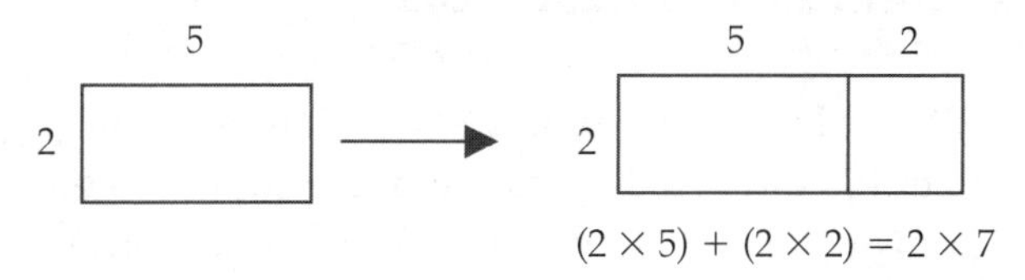

$(2 \times 5) + (2 \times 2) = 2 \times 7$

2. Doubling and halving can be examined and then the array can be used as a tool to explore its generalization to the associative property:

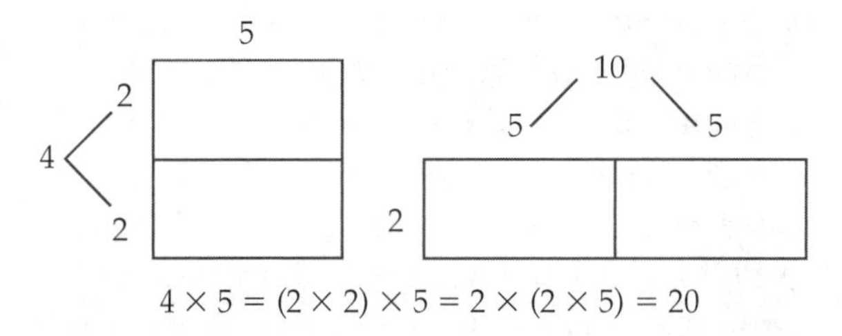

$4 \times 5 = (2 \times 2) \times 5 = 2 \times (2 \times 5) = 20$

3. The commutative property, equivalence, and congruence:

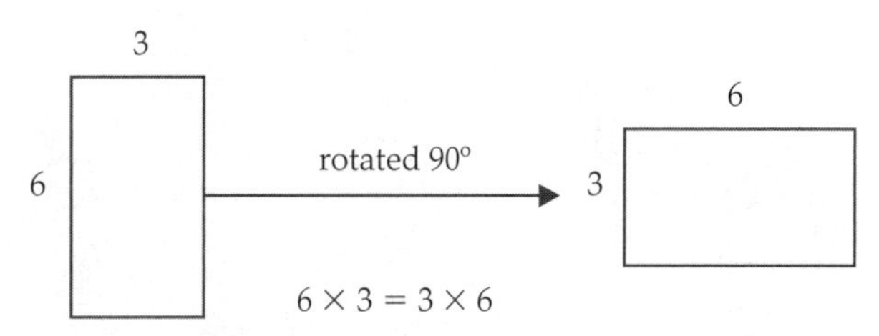

$6 \times 3 = 3 \times 6$

Many opportunities to discuss these landmarks in mathematical development will arise as you work through this unit. Look for moments of puzzlement. Don't hesitate to let students discuss their ideas and check and recheck their strategies. Celebrate their accomplishments! They are young mathematicians at work!

A graphic of the full landscape of learning for multiplication and division is provided on page 11. The purpose of the graphic is to allow you to see the longer journey of students' mathematical development and to place your work with this unit within the scope of this long-term development. You may also find the graphic helpful as a way to record the progress of individual students for yourself. Each landmark can be shaded in as you find evidence in a student's work and in what the student says—evidence that a landmark strategy, big idea, or way of modeling has been constructed. In a sense, you will be recording the individual pathways your students take as they develop as young mathematicians!

References and Resources

Battista, Michael T., Douglas H. Clements, Judy Arnoff, Kathryn Battista, and **Caroline Auken Bomsn**. 1998. Students' spatial structuring of 2D arrays of squares. *Journal for Research in Mathematics Education* 29:503–32.

Beishuizen, Meindert. 1993. Mental strategies and materials or models for addition and subtraction up to 100 in Dutch second grades. *Journal for Research in Mathematics Education* 24:294–323.

Dolk, Maarten, and **Catherine Twomey Fosnot**. 2005a. *Fostering Children's Mathematical Development, Grades 3–5: The Landscape of Learning.* CD-ROM with accompanying facilitator's guide by Sherrin B. Hersch, Catherine Twomey Fosnot, and Antonia Cameron. Portsmouth, NH: Heinemann.

———. 2005b. *Multiplication and Division Minilessons, Grades 3–5.* CD-ROMs with accompanying facilitator's guide by Antonia Cameron, Carol Mosesson Teig, Sherrin B. Hersch, and Catherine Twomey Fosnot. Portsmouth, NH: Heinemann.

Fosnot, Catherine Twomey, and **Maarten Dolk**. 2002. *Young Mathematicians at Work: Constructing Multiplication and Division.* Portsmouth, NH: Heinemann.

Gravemeijer, Koeno P. E. 1991. An instruction-theoretical reflection on the use of manipulatives. In *Realistic Mathematics Education in Primary School,* ed. Leen Streefland. Utrecht, Netherlands: Freudenthal Institute.

———. 1999. How emergent models may foster the constitution of formal mathematics. *Mathematical Thinking and Learning* 1 (2): 155–77.

Karlin, Samuel. 1983. Eleventh R. A. Fisher Memorial Lecture. Royal Society, 20 April.

Klein, Anton S., Meindert Beishuizen, and **Adri Treffers**. 2002. The empty number line in Dutch second grade. In *Lessons Learned from Research,* ed. Judith Sowder and Bonnie Schapelle. Reston, VA: National Council of Teachers of Mathematics.

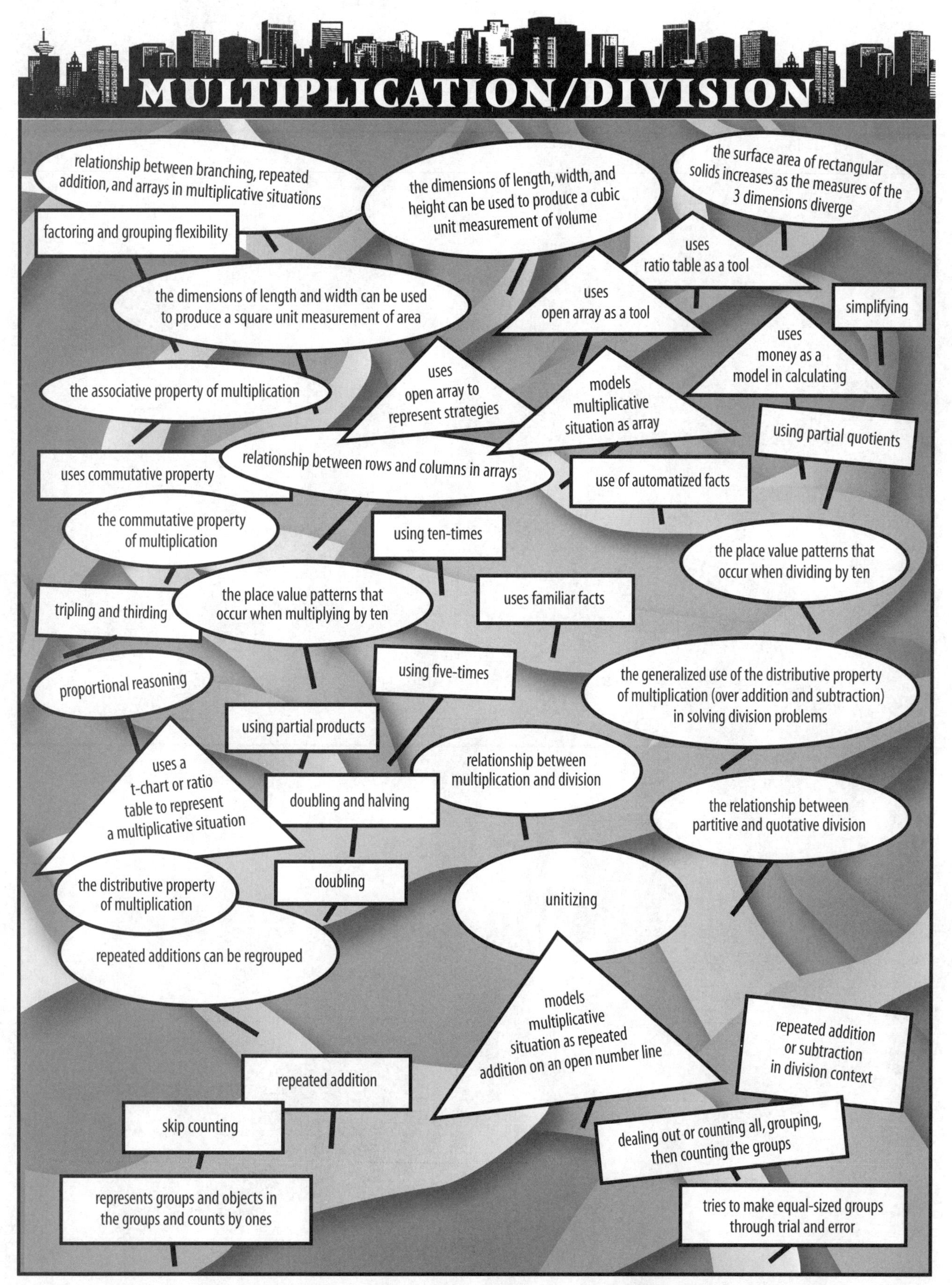

The landscape of learning: multiplication and division on the horizon showing landmark strategies (rectangles), big ideas (ovals), and models (triangles).

DAY ONE

Muffles' Truffles

Today you will be developing the context, laying the terrain for the work of the next two days. After you introduce the situation and show students the poster of Muffles' Truffles shop, students will set to work figuring out how many boxes will be needed for various quantities of truffles and how many pieces of candy will be left over to make mixed assortments. As you move around the room supporting, encouraging, and conferring, you will notice that students use a variety of strategies as they work on this investigation. After their explorations, students make posters of some of the ideas they want to share in a math congress, to be held on Day Two.

Day One Outline

Developing the Context

- Tell the story of Muffles' truffles boxes and discuss how many truffles would fit in a 2 × 5 box.
- Ask students to work on the problems in Appendix B.

Supporting the Investigation

- Note students' strategies and encourage them to consider ways to calculate more efficiently.

Preparing for the Math Congress

- Ask students to make posters of their strategies.
- Plan to focus the congress discussion on unitizing, multiplication and division by ten, and connections among students' strategies.

Materials Needed

Muffles' Truffles shop poster [If you do not have the full-color poster (available from Heinemann), you can use the smaller black-and-white version in Appendix A.]

The truffles investigation sheet (Appendix B)—one per pair of students

Drawing paper—a few sheets per pair of students

Large chart paper—at least one sheet per pair of students

Connecting cubes, as needed

Large chart pad and easel (or chalkboard or whiteboard)

Markers

Developing the Context

- Tell the story of Muffles' truffles boxes and discuss how many truffles would fit in a 2 x 5 box.
- Ask students to work on the problems in Appendix B.

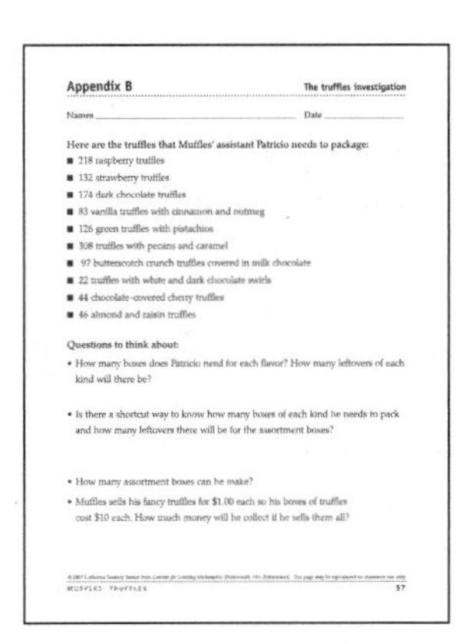

Appendix B — **The truffles investigation**

Names ____________ Date ________

Here are the truffles that Muffles' assistant Patricio needs to package:

- 218 raspberry truffles
- 132 strawberry truffles
- 174 dark chocolate truffles
- 83 vanilla truffles with cinnamon and nutmeg
- 126 green truffles with pistachios
- 308 truffles with pecans and caramel
- 97 butterscotch crunch truffles covered in milk chocolate
- 22 truffles with white and dark chocolate swirls
- 44 chocolate-covered cherry truffles
- 46 almond and raisin truffles

Questions to think about:

- How many boxes does Patricio need for each flavor? How many leftovers of each kind will there be?
- Is there a shortcut way to know how many boxes of each kind he needs to pack and how many leftovers there will be for the assortment boxes?
- How many assortment boxes can he make?
- Muffles sells his fancy truffles for $1.00 each so his boxes of truffles cost $10 each. How much money will he collect if he sells them all?

MUFFLES' TRUFFLES 57

Display the truffles shop poster (or Appendix A) as you tell the following story:

> *Muffles has a small truffles shop. He makes truffles and packages them in boxes of ten. When he first opened his shop, he had only a few customers—his family and friends. His truffles were so delicious, so delectable, that soon his customers couldn't stop eating them. They also couldn't stop talking about the most delicious truffles in the world. They told their friends, who in turn told their friends, who in turn told their friends, and before Muffles knew it he had so many customers he could hardly keep up with the demand for truffles. Long lines of people waited outside his door; sometimes the line snaked around the corner. Sometimes there were so many customers that Muffles ran out of truffles. What sad faces! What disappointment!*
>
> *Muffles tried to keep up. He got up very early in the morning. He worked late into the night. But no matter what he did, he just couldn't keep up with the demand for truffles. Finally, Muffles had an idea. He hired assistants to help him. Now one assistant packages the truffles and the other sells them. All Muffles has to do is make the truffles, which is what he really, truly loves to do.*
>
> *Muffles is happy again. He loves getting up early in the morning to make his candies. After he makes a batch of a certain kind of truffle, he counts them, puts them on a tray, and labels the tray for his assistant, Patricio, who will package them. When Patricio arrives later in the morning, he has many trays or batches of many kinds of truffles to package. He has strawberry truffles, dark chocolate truffles, truffles with pecans and caramel, and raspberry truffles. He has green truffles with pistachios, and vanilla truffles with cinnamon and nutmeg. So many kinds, it's no wonder Muffles is becoming famous! When Patricio sees all the trays, he thinks, "How many boxes do I need to get from the shelf? Will there be any leftovers that can be used to make boxes that might contain an assortment?" This is what each box looks like.*

Pause at this point in the story and draw the box as a 2×5 grid, like this:

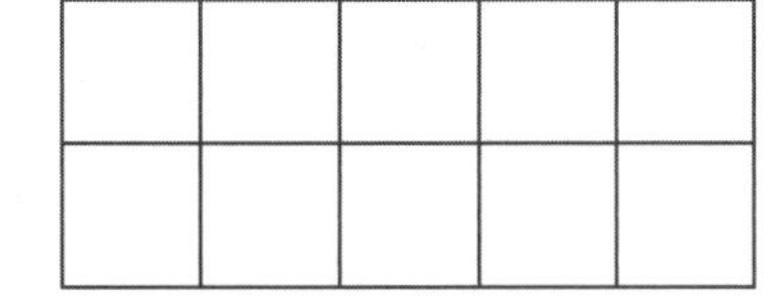

Ask students,

- *One truffle fits in each square. How many truffles would this box hold?*
- *How do you know?*

Ask students to describe how they know and listen for whether they add five and five, notice two rows of five or five columns of two, or need to count by ones, etc. Acknowledge all the ways to think about the box and then label it 2×5, for two rows and five columns. Then draw a rectangle and write 10 on it, like this:

10

Now go on with the story.

One day Muffles makes ten different kinds of truffles. This is the list of the different trays Patricio sees when he comes into work that day:

- *218 raspberry truffles*
- *132 strawberry truffles*
- *174 dark chocolate truffles*
- *83 vanilla truffles with cinnamon and nutmeg*
- *126 green truffles with pistachios*
- *308 truffles with pecans and caramel*
- *97 butterscotch crunch truffles covered in milk chocolate*
- *22 truffles with white and dark chocolate swirls*
- *44 chocolate-covered cherry truffles*
- *46 almond and raisin truffles*

Behind the Numbers

The total number of boxes that can be made, including the assortment boxes, is 125. The numbers have been chosen very carefully to allow for various levels of mathematizing. The units (for the assortment boxes) as well as the tens (for the boxes of one kind of truffle) can be cleverly grouped into tens and multiples of tens. The question about the money is also important as it allows you to notice if students can think reversibly. For example, once they have determined that 8 boxes will be used for 83 vanilla truffles, do they realize that the cost of the 8 boxes will be $80? Once they determine that a total of 125 boxes will be needed, do they multiply by 10 to calculate the money ($1250), skip-count, or perhaps even think this is a new, unrelated problem?

Ask students to investigate the following:

- *How many boxes does Patricio need for each flavor? How many leftovers of each kind will there be?*
- *Is there a shortcut way to know how many boxes of each kind he needs to pack and how many leftovers there will be for the assortment boxes?*
- *How many assortment boxes can he make?*
- *Muffles sells his fancy truffles for $1.00 each so his boxes of truffles cost $10 each. How much money will he collect if he sells them all?*

Facilitate a brief preliminary discussion in the meeting area before the students set to work, just to ensure that they understand the investigation. Assign math partners and distribute a copy of Appendix B to each pair of students. Have drawing paper and bins of connecting cubes available for students to use if they wish.

Supporting the Investigation

☀ Note students' strategies and encourage them to consider ways to calculate more efficiently.

As students work, walk around and take note of the strategies you see. Confer with students as needed to support and challenge them. Do *not* show them what to do; instead, note their strategies and look for moments when you can encourage reflection and, if appropriate, puzzlement.

Here are some strategies you might see for determining the number of boxes:

- Drawing a representation of the situation: making boxes of truffles for each batch and counting the truffles by ones as they are filling the box, then counting and labeling the boxes
- Drawing boxes and labeling them with 10 (instead of drawing each truffle), then skip-counting *[See Figure 1]*
- Using additive reasoning (expanded notation): $132 = 100 + 30 + 2$; 100 is 10 boxes, 30 is 3 more, so that's 13 boxes and 2 truffles left over *[See Figure 2]*
- Combining the units for the assortments to make tens—for example, 3 and 7, 4 and 6, etc.—but not realizing that this is possible with the number of tens as well (other students may just add up all the units in a procedural fashion)
- Unitizing and using place value—knowing that the number of boxes and candies left over is represented by the numbers; grouping the tens together in friendly ways *[See Figure 3]*

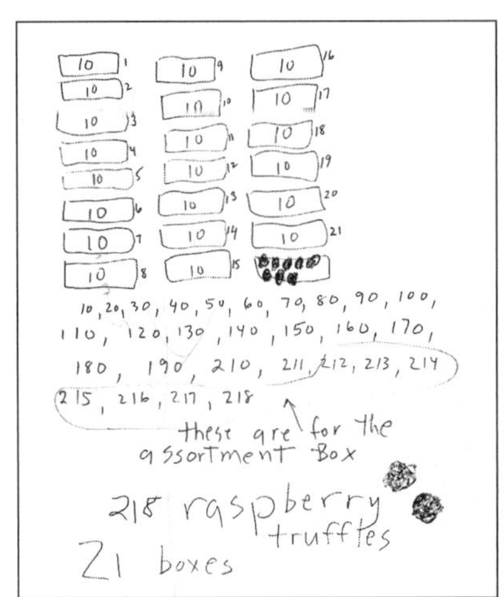

Figure 1

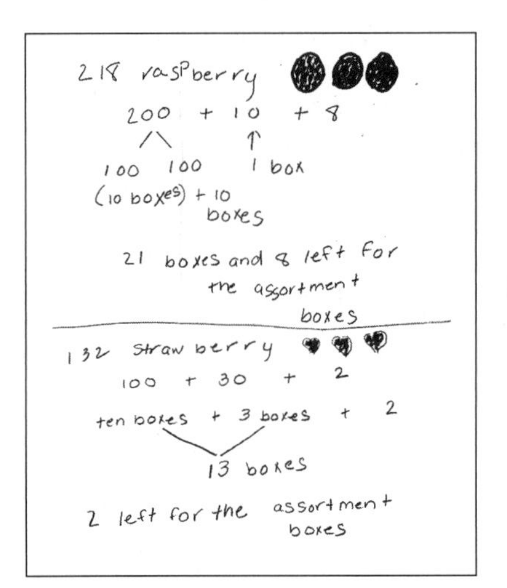

Figure 2

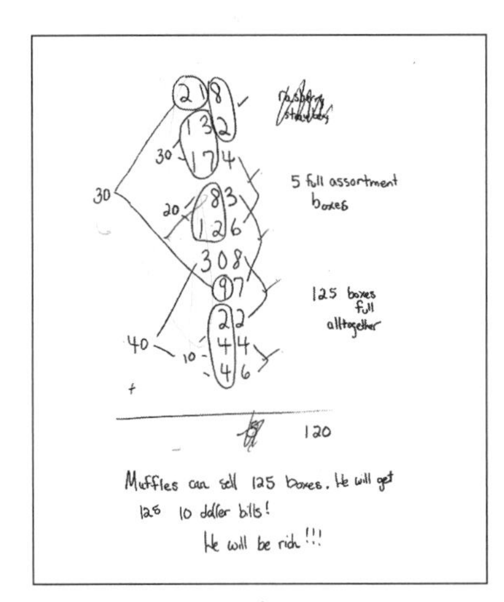

Figure 3

Here are some strategies you might see as students determine how much money Muffles will collect:

- Counting by ones
- Using the context of the boxes and skip-counting by tens
- Adding the amounts to get a total of the truffles, not making use of the boxes of ten
- Counting the number of boxes and multiplying by ten

As you confer, listen first and try to understand what the student is attempting to do. Supporting that idea, even if its execution would be tedious, is the starting place of every conference. Then look for ways to encourage and challenge the student that will support the development of greater efficiency. For example, if you see students drawing every truffle, be sure to help them think about whether they need to do this. Questions like "That looks like a lot of work—do you have to draw the box like this [with the truffles inside] or is there a faster way to show that each box holds ten?" or "Do you need to draw boxes?" may help students move away from inefficient strategies. You may find, however, that some students do need to draw every box and a few may even need to use cubes to act out the situation. Remember to work with the mathematician; don't just try to fix the mathematics!

Differentiating Instruction

Students with a strong understanding of place value may quickly realize that for each quantity of truffles, the digits to the left of the ones are the number of boxes of ten and the ones digit will be the number of candies left over. But how deeply do they understand this big idea? Have they only generalized it to two-digit numbers?

To probe students' understanding of place value, ask them to think about numbers in the hundreds. Students who quickly know the answer for double-digit amounts may need time to explore greater quantities. This will be especially true for students who have a rote understanding of place value and may have been taught to identify ones, tens, and hundreds in columns as an activity to fill in place value charts. They may be quite surprised to see that 126 truffles make 12 boxes and that they do not necessarily need to stop to think about the hundred being 10 tens in order to calculate this. It is important to help students realize that the amount 126 is not just 1 hundred, 2 tens, plus 6 ones: it is also 12 tens plus 6 ones. It is important to connect this understanding to multiplication by ten: $12 \times 10 = 120$. Sometimes, as students explore the context of large quantities of truffles, they also begin to think about the way numbers in the hundreds are related (i.e., if there are 10 boxes of 10 in 100, then in 200, there are 20 boxes of 10). Pose questions to help students generalize unitizing, and division and multiplication by 10. At the same time, providing opportunities to group the truffles into tens both with manipulatives and with written representations (for those whose understanding of place value is very weak) is an important way to differentiate this lesson for the range of learners in the classroom.

Behind the Numbers

Although students may use additive reasoning to solve this problem, the mathematical focal point should be the multiplicative structure of our base-ten system, and quotative division. With 132 truffles, the question of how many boxes of ten can be made and how many candies will be left over is, in essence, asking students to think about how many tens fit into 132. By nature, measurement (quotative) division problems are easier for students than fair-sharing (partitive) problems because they know what is in the group (in this instance, the group is a box of ten) and they can use this quantity to reason with, usually using a strategy such as repeated addition. This is one reason that skip-counting is a common beginning multiplication and division strategy. Be aware, however, that students' understanding of place value may mask the division that is actually occurring. For example, that $132 \div 10$ is 13 remainder 2, or that 132 is also equivalent to $13 \times 10 + 2$ may not be explicit in students' solutions. This is another reason why the money question is important. It provides another opportunity to revisit multiplication by ten and the place value pattern that results.

Inside One Classroom

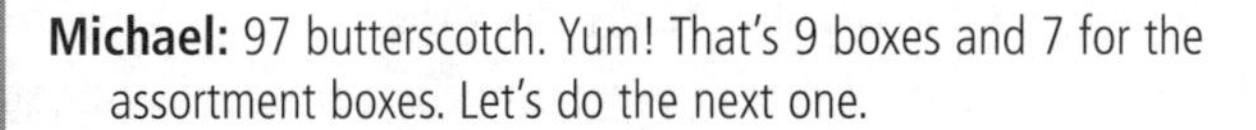

Conferring with Students at Work

Michael: 97 butterscotch. Yum! That's 9 boxes and 7 for the assortment boxes. Let's do the next one.

Toni (the teacher)**:** Tell me about your strategy. How did you know that so fast?

Callie: There's 9 in the tens column. See? *(Points to the 9.)*

Toni: Oh. That's helpful, isn't it? What about the 132 strawberry truffles? Will there be 3 boxes?

continued on next page

Author's Notes

Toni begins by listening and observing to see how Michael and Callie are mathematizing this situation. She notices they are using place value and begins the conference by questioning to see how deep this knowledge goes.

continued from previous page

Michael: I don't think it works for big numbers. That's 1 hundred, and 3 tens, and 2 ones...let's...hey, the 100 is 10 boxes. Three more...that's 13 boxes and 2 left for the assortment boxes.

Toni: That's interesting—132 truffles make 13 boxes and 2 left. Will that pattern happen again?

Michael: Maybe, I don't know. Let's try 126. Maybe it will be 12.

Toni: If you are right and it is 12, how much money will Muffles collect from those 12 boxes? And how much from these 13 boxes? How could we figure that out?

Callie: Count by 10? *(Counts and uses her fingers to keep track.)* So it's 10, 20, 30, 40...100, 110, 120.

Michael: Hey! It is 12 boxes. Because that's 120 and the 6 leftovers make 126!

Toni: And the 13 boxes? How many truffles is that again? I forget.

Michael: It's 130! Yeah, 13 boxes is 130.

Toni: So it seems like you have a shortcut way of figuring this out. Do you think you could figure out a shortcut to know the amount of money, too—so you wouldn't need to skip-count? When we prepare for the math congress later, that would be a nice idea to come to the community with...an explanation of what you noticed and a proposal for what Muffles' assistant, Patricio, could do. But you'll need to convince us. So think about how to explain it.

By asking the students to think about the number of boxes in relation to the number of truffles, Toni is pushing one of the big ideas in place value: unitizing. Learning to think about a group of ten as one unit, while simultaneously thinking about the objects in the group as one unit, is a struggle students often have in the construction of place value.

Toni challenges the students to consider the money. This pushes them to consider multiplication by ten and its connection to place value: 12 x 10 = 120.

Toni stays grounded in the context to help students realize the meaning of what they are doing. She does not talk about tens, but about boxes and truffles, and she focuses on multiplication by ten, not on the rote reading of the digits in the columns.

Toni leaves the conference with an invitation to the young mathematicians to build a convincing argument for their shortcut. Now they will need to examine the place value patterns that result when multiplying or dividing by the base, and be sure of their thinking in order to convince others.

Assessment Tips

Moving around the room and conferring with students as they worked gave you the opportunity to notice how the students were thinking about the work, and what struggles they were having. As they worked, you might have noticed students grappling with unitizing as they considered the number of truffles in relationship to the number of boxes of ten that could be made from them. Did students need to draw boxes? If so, how did they count? Did they use repeated addition? Skip-count? Find clever ways to pair amounts to make groups of ten? Did some strategies evolve from drawing or skip-counting when the students discovered the relationships among the numbers of boxes of ten, the truffles left over, and the total number of truffles? Did some students quickly unitize and recognize that the number of boxes and leftovers is already represented by the digits in the columns of the original number if one uses the place value? If so, how did they generalize this idea? Could they generalize it for other, greater numbers—for example, the 125 boxes bringing in $1250 if they sell for $10 each? How did students organize their work? Did they find systematic ways to communicate their ideas? If they drew the boxes, did they realize that each truffle was in a row and a column simultaneously?

These are all important observations to keep in mind—they tell you where your students are as they begin the journey with this unit. Take some notes, date them, and place these in student portfolios.

Preparing for the Math Congress

As students finish their work, ask them to make posters of their strategies for display. The purpose of doing this is to foster further reflection. Often it is only when students make posters that their ideas really become solidified. Reflecting on the important things they want to explain, deciding what is extraneous, discussing how to revise their draft work, and determining how to represent the main issues provide time for thinking about their thinking: metacognition. Encourage students to be clear on their posters so their classmates will be able to read and understand what they did. When mathematicians write up their mathematics for math journals, they do not merely reiterate everything they did. They craft a proof or argument for other mathematicians. Doing this not only generates further reflection, it focuses the author on developing a convincing and elegant argument—an important part of mathematics. Of course, elementary students will not be expected to write formal proofs, but by focusing on the justification and logic of their arguments you are helping them develop the ability to write up their ideas for presentation in a mathematical community. For example, if students have constructed the idea that one shortcut way of knowing the number of boxes and leftovers is just to use the digits in the given quantity, push them to generalize the idea, and suggest that they focus their poster on justifying this generalization.

- Ask students to make posters of their strategies.
- Plan to focus the congress discussion on unitizing, multiplication and division by ten, and connections among students' strategies.

Tips for Structuring the Math Congress

As the students prepare their posters, take note of the different strategies you see. The math congress will not be held until Day Two. The purpose of the congress is to support students in considering unitizing, the place value patterns when multiplying or dividing by ten, and a variety of strategies and how they are connected. Students should also examine the strategies for efficiency.

Plan on structuring a congress to discuss unitizing and place value—the fact that the meaning of the digits in a number is connected to groupings of ten. You will probably have ample student work illustrating this idea. Also plan on helping students understand the connection of multiplication by ten (the money question) to division by ten (the number of boxes that can be made, with ten in a box) and the resulting place value patterns: $12 \times 10 = 120$.

Reflections on the Day

Students were introduced to the array model today within the context of Muffles' boxes—2×5 arrays. They explored place value, and multiplication and division by ten. On Day Two they will examine these ideas more explicitly in a math congress.

DAY TWO

Muffles' Truffles

Materials Needed

Students' posters from Day One

Before class, display all the posters around the meeting area.

Sticky notes—one pad per student

Large chart pad and easel

Markers

Today begins with a "gallery walk," with students walking around and examining the posters from Day One and posting comments and questions on them. A math congress then convenes to discuss a few of the strategies displayed on the posters, and to consider big ideas related to place value and to multiplication and division by ten. The congress gives you a chance to focus on a few of the big ideas and strategies that came up in the students' work that you think will be helpful for the entire class to examine and discuss.

Day Two Outline

Preparing for the Math Congress (continued from Day One)

- Conduct a gallery walk to give students a chance to review and comment on each other's posters.

Facilitating the Math Congress

- Focus the congress discussion on the posters you selected on Day One.

Preparing for the Math Congress (continued from Day One)

Explain to students that they will have a gallery walk to look at the posters before you start a congress. Hand out small pads of sticky notes and suggest that students use them to record comments or questions. These notes can be placed directly on the posters. Display all the posters around the room and have everyone spend about fifteen minutes walking around, reading and commenting on the mathematics on the posters. Give students a few minutes to read the comments and questions posted on their own work and then convene a math congress.

Conduct a gallery walk to give students a chance to review and comment on each other's posters.

If students have never participated in a gallery walk, expect that their comments and questions may be trivial. With experience, students will develop the ability to pose relevant questions and understand what kinds of comments are helpful to the creators of the posters. You may have to spend time examining the kinds of questions and comments made by students on a given poster in order to help them understand what constitutes an appropriate response.

Facilitating the Math Congress

Gather the students in the meeting area. Have them sit next to their partners. Focus conversation on two or three posters and scaffold a conversation that will facilitate development of the community. Begin by inviting one pair of students to display their work and share their strategy. Invite questions and comments before you go to a second piece of work.

Focus the congress discussion on the posters you selected on Day One.

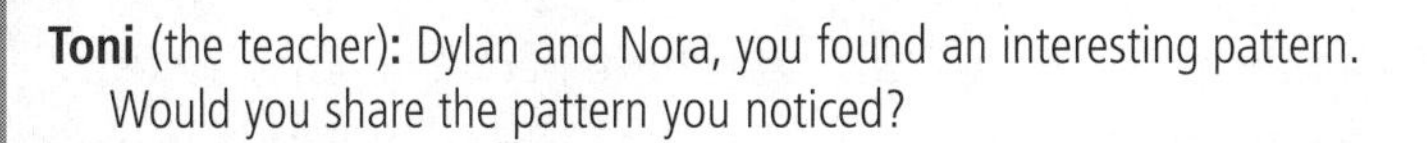

A Portion of the Math Congress

Toni (the teacher)**:** Dylan and Nora, you found an interesting pattern. Would you share the pattern you noticed?

Nora: We noticed that the number of boxes and the number of leftovers was right there...in the number of truffles! With 53 truffles, the number of boxes is 5 and the ones left for the assortments is 3.

Dylan: First we didn't think it would work with the bigger numbers, the ones more than 100. But it did! Because 126 truffles is 12 boxes and 6 left for the assortments.

Mary: We saw that, too. But why is that happening? That's what I want to know.

Nora: When we did the money...we realized why. Muffles will make $120 on 12 boxes...and that's because he is selling 120 truffles: 12 × 10.

continued on next page

Author's Notes

Toni starts by asking two students to share the generalization they have been working to prove. This move implicitly suggests that part of doing mathematics is communicating and justifying thinking to a community of other mathematicians.

Discussion is welcomed.

continued from previous page

Toni: A lot has been said so far. Does anyone have a comment or a question about any ideas that have been presented to us? No? Could someone put in your own words what Dylan and Nora are trying to convince us of?

Asking for clarification and paraphrasing ensures that students understand each other's ideas and can discuss them.

John: I have a question. Don't 12 and 6 add up to 18? I don't get what they're talking about—12 and 6 isn't 126.

Dylan: It's not really 12 and 6. It's 12 × 10 plus 6. The 12 boxes are 12 tens. There's 10 truffles in a box. Oh—this is so hard to explain.

Kareem: I think I understand. The 12 isn't just a 12. It's 12 groups of 10. That's 120.

Students, not the teacher, defend their thinking.

Toni: I'm wondering what all these boxes would look like on the tray. Let's try to imagine it. *(On chart paper draws 12 boxes—6 rows of boxes, 2 boxes in each row—to make a 12 × 10 array of truffles.)* Is this 120? Check me. And do we have 12 boxes?

Dylan: *(Counting the boxes.)* Yep. Hey. It's also 12 rows of truffles and each row has 10. That's cool.

Toni seizes the moment to emphasize the unitizing underlying Kareem's statement, and to make explicit the multiplicative nature of the place value system. She also stays grounded in the context and makes an array of the total amount.

Toni: So one way of thinking about the number 126 is this way—and I'm going to use Kareem's words, but record it the way mathematicians would. *(Writes on the board:)*

$(12 \times 10) + 6$

Twelve groups of 10 and 6 more. Questions? Comments?

Assessment Tips

It would be nice to place the posters in students' portfolios; however, they are probably too large. If so, you can take a photograph of each poster and staple it to a blank page for your anecdotal notes. Make notes about the strategies and big ideas described in the unit overview (pages 5–11). Do you have evidence that any of these ideas and strategies have been constructed?

Over the course of the unit, you might want to keep all student work together. This work can be examined with students on Day Ten as a way to reflect on their learning of the past two weeks. This work can then be used to create a sociohistorical wall for reflection.

Reflections on the Day

Today students examined the place value in our number system. They were supported to develop the idea of unitizing and its application to place value. They also examined multiplication and division by ten. At this point their work is grounded in the context of boxes with rows and columns (2 × 5). As the unit progresses, boxes having different configurations of candy will be introduced to deepen students' understanding of arrays.

DAY THREE

Designing More Boxes

Today's math workshop begins with a minilesson to remind students of the work they did on Day Two when they examined multiplication and division by ten. After the minilesson, students begin a different investigation designing a new set of boxes for Muffles.

Day Three Outline

Minilesson: Around the Circle

- Facilitate a minilesson focused on multiplication and division by ten and the place value patterns that result.

Developing the Context

- Introduce a new truffles box scenario and ask students to create new box designs and calculate the price of each box.

Supporting the Investigation

- Encourage students to consider how the configuration of one box might be used to help with another.

Materials Needed

Chart-size graph paper (with one-inch-squares)—several sheets per pair of students

Drawing paper—one sheet per pair of students

Scissors—one pair per student

Large chart pad and easel

Markers

Minilesson: Around the Circle (10–15 minutes)

Facilitate a minilesson focused on multiplication and division by ten and the place value patterns that result.

This mental-math minilesson is designed to encourage students to use the place value patterns when multiplying by ten. Have the students sit in a circle. Ask them to count around the circle aloud by ones and write their numbers on a piece of paper. Then ask a student to display his or her number for everyone to see. Explain that now everyone will count around the circle again but this time counting by tens; with the student who said "one" the first time around saying "ten" this time. Ask for predictions of what number the students will get this time. [Note: For example, a student who wrote down 16 last time should get 160 this time.] Do this several times with different numbers. If guesses are random, stop after a few students have skip-counted and ask students to reconsider their original predictions: would they change these, and why? If this procedure is easy for the students, count by ones around the circle twice or three times so that greater numbers come up. At the end of the count-around, highlight the mathematics in the following way: "Mary had a 23; then we counted by tens and she had 230. Nora had a 48 (after counting around twice). When we counted by tens, she had 480." You might also write $23 \times 10 = 230$. Discuss the place value pattern that students notice and ask if they think the pattern is connected to the work they did with the boxes on Day Two.

Inside One Classroom

A Portion of the Minilesson

Toni (the teacher): *(Records students' predictions on chart paper as they say them. The number was 32.)* OK, 120, 200, 300, 322, 480, 200, 260, 240. Any other predictions? OK. Are there any predictions up here that you feel are unreasonable, given that we have 32 students counting and we're counting by 10?

Giles: Yeah, 120.

Toni: Why 120?

Giles: It seems too low. When 10 kids count, we would already be at 100. So 32 kids counting has to be a lot more than 120.

Toni: What do you think about what Giles just said? Any other predictions that you feel are unreasonable?

Katie: I don't know how unreasonable it is, but I know we won't end at 322.

Toni: And how do you know that?

continued on next page

Author's Notes

Toni records all predictions, but then encourages students to think about the reasonableness of their estimates.

Rather than agreeing or disagreeing with Giles, Toni poses a question. This neutrality supports both Giles, who gives the answer (he needs to explain or justify his thinking to the community), and the other students in the class (they need to consider why 120 might be an unreasonable prediction).

continued from previous page

Katie: Because the number has to end in 0.

Toni: Why is that?

Katie: Because that's what happens when you count by 10. Look, 10, 20, 30, 40… You're not going to say any number that doesn't end in 0.

Dylan: It's going to be 320. It's just like Muffles' boxes yesterday. It will be 32 tens.

Toni: Hmmm. So it sounds like you're making a prediction. *(Writes on the board: "Katie's conjecture: When we count by 10, all the numbers will end in 0. Dylan's conjecture: The 32 just moves over. It's 32 tens like Muffles' boxes.")* How many people agree with Katie and Dylan? Well, let's see what happens when we count. I'll record what people say on the board.

Toni looks for important moments when generalizing might occur. That all multiples of ten end in zero and that the number just moves over a place are two big ideas for students to develop. Rather than commenting on a generalization or trying to get all students to understand or agree with it at that moment, she poses it as an idea that the community can consider as they do the count-around. Although these ideas were discussed on Day Two, some students in the class need opportunities to ponder them again.

Developing the Context

After the minilesson, tell the students about Muffles' new predicament:

> *One day a customer comes into Muffles' shop and asks him why he only packages his truffles in boxes of ten, in two rows with five columns. "What if I want to buy more or fewer? Don't you have other sizes of boxes?" This question gets Muffles wondering: what other sizes of boxes could he make? What should he charge for them?*

Introduce a new truffles box scenario and ask students to create new box designs and calculate the price of each box.

Ask students to work in pairs to design rectangular boxes (just one layer) in other sizes (all smaller than 10×10). Have students use graph paper with one-inch squares and cut out the rectangles as blueprints for Muffles. On the front they should record the numbers of rows and columns, for example 2×6 for 2 rows and 6 columns. Have them calculate the price of each box, reminding them that the truffles cost $1 each, and write the price on the back.

Behind the Numbers

These graph paper arrays can be used to develop fluency with the basic multiplication facts. As students work to calculate the price of each box, have them explore relationships between the boxes. For example, if they know a 5×5 box of truffles costs $25, they can lay this over a 6×5 box of truffles to realize that the larger box costs just $5 more. As you encourage the students to explore the relationships among their boxes, you are also supporting the development of the distributive property, showing that $6 \times 5 = (5 \times 5) + (1 \times 5)$; the commutative property, showing that $2 \times 6 = 6 \times 2$ (representing the box being turned); and the associative property, showing that $(2 \times 2) \times 6 = 2 \times (2 \times 6)$; a 4×6 box is the same price as a 2×12 one, as a 2×6 section has just been moved.

Supporting the Investigation

Encourage students to consider how the configuration of one box might be used to help with another.

Assign math partners and give each pair of students a large, chart-size piece of graph paper and scissors. As students work, walk around and take note of the strategies you see. Confer with students as needed to support and challenge their investigation. Focus not on the answer but on the thinking—on the strategies that students are trying out. Support them in thinking flexibly by encouraging them to consider how the configuration of one box might be used to help with that of another.

If the idea of box congruence comes up initially, explore it. Even after much exploration, some students may not be convinced that the rectangles represented by the bottoms of some boxes may be congruent. For students who recognize that the rectangles are the same, encourage them to consider how they might *convince* other members of the mathematical community about this idea. Challenged in this way, students often rotate the array to demonstrate congruence. Here they are providing a justification for the commutative property of multiplication:

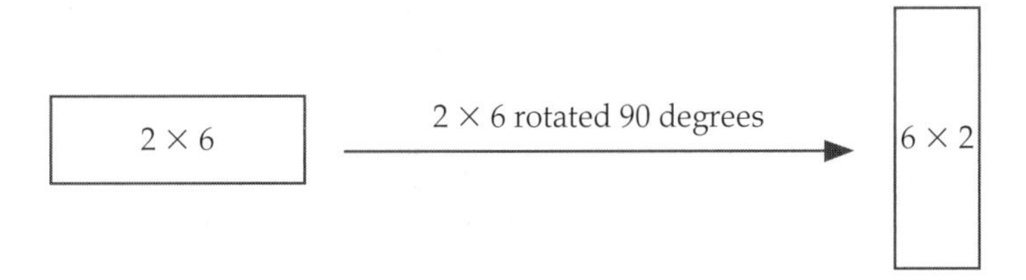

On the other hand, the rectangles represented by the bottoms of some boxes will be equivalent in area, but not congruent. For example, a 2 × 6 box and a 4 × 3 box both hold 12 truffles. One dimension has been halved; the other, doubled. The boxes are equivalent but not congruent. Don't shy away from introducing these terms as the need for them arises. Introducing mathematical terminology in connection with a story context is an appropriate and effective method.

Assessment Tips

Watch how students work. When they determine the size of the array, what do they do with the square in the corner? Often students think that since they counted it once when they were counting the number of rows, they should not count it again when they count the number of columns. Helping students realize that each square on the graph paper (each truffle) is in a row and a column simultaneously is very important. It is a big idea underlying students' emerging understanding of arrays. There is a difference between counting each truffle (or each square) one-by-one and counting rows and columns. In the latter case students have organized the total into an arrangement of repeated groups.

Take note of students' strategies for determining the price. Do they just cut out arrays and then count each square one by one? Do they skip-count or use repeated addition? Do they anticipate the price by adding partial products—for example, calculating 6×7 by using $(6 \times 4) + (6 \times 3)$? Are they using doubling and halving strategies? How do students use their knowledge of the number system? For example, do they say things like "seven times five can't be 34 because when you count by five, you wouldn't end on a number that has a four" or "Since 24 is an even number, I can make a box for 24 truffles with two rows"? As you move from group to group, be sure to keep track of these kinds of comments. The notes you take now will also help you plan future discussions to support students as they continue to explore arrays.

Inside One Classroom

Conferring with Students at Work

Author's Notes

Toni (the teacher)**:** I see that you are cutting out boxes and then counting each square to figure out the price. I wonder if this box that you just did, the 5×5, could be used to help you figure out this 5×4? *(Lays one array on top of another.)*

Toni notices that some students are still counting the squares by ones. By laying one array on top of another, she encourages them to begin to examine relationships between the arrays.

Chris: Oh. It's just $5 less. Twenty-five minus 5...that's 20. Oh, yeah, that's easier.

Toni: Well, that's a good thing to notice. That does make it easier, doesn't it? It saves a lot of counting. And mathematicians like to find efficient ways, don't they? Maybe in the math congress later you can share the shortcuts you find...the way you use relationships between the boxes to calculate the prices.

Toni invites students to present their findings in the math congress. Implicitly she is establishing what holds up as important mathematical behavior, and what counts as an idea worth discussing.

Reflections on the Day

The minilesson today encouraged students to continue thinking about how multiplying a given number by ten has the effect of moving the number over a place. That is, the digits all shift to the next place value position. The investigation of designing new boxes for Muffles provided students with opportunities to continue exploring arrays and the relationships among them. Words like *congruent* and *equivalent* were introduced in the context of student activity, and students began explorations of the commutative, associative, and distributive properties. In the math congress on Day Four, students will have a chance to explore these ideas more deeply.

DAY FOUR

Designing More Boxes

Materials Needed

Students' work from Day Three

Chart-size graph paper (with one-inch squares)—several sheets per pair of students

Scissors—one pair per student

Large chart paper—one sheet per pair of students

Large chart pad and easel

Markers

Students are given time to complete their new box designs from Day Three and make posters of the relationships they noticed, including the helpful strategies they found to calculate the prices. A math congress is then held to discuss these ideas.

Day Four Outline

Preparing for the Math Congress

- Give students time to complete their box designs and then have them prepare posters of their strategies.
- Plan to focus the congress on the commutative and associative properties and the difference between congruence and equivalence.

Facilitating the Math Congress

- Have a few students share how they used the price of one box to help calculate the price of another and push for generalization of that strategy.
- Ask students to use the graph paper arrays to illustrate their strategies. Record their box designs as open arrays.

Preparing for the Math Congress

Convene students in the meeting area and explain that they will have time today to finish their new box designs for Muffles. Then they will meet for a math congress to discuss the relationships they noticed among the boxes and the interesting strategies and shortcuts they used to calculate the price of each new box. Have students consider some general statements they can make about how to use the information they have about one box to help with another. Ask them to prepare posters for the math congress. As students work, notice the strategies and ideas they think are important to share in the congress.

- Give students time to complete their box designs and then have them prepare posters of their strategies.
- Plan to focus the congress on the commutative and associative properties and the difference between congruence and equivalence.

Tips for Structuring the Math Congress

You will want to focus the math congress on the distributive, commutative, and associative properties. Look for samples of work that demonstrate these properties. Students may not have generalized the properties yet, but several samples of student work will probably show specific cases of their use. It is also important to look for work that will allow you to discuss the difference between congruence and equivalence.

Facilitating the Math Congress

After students have had a sufficient amount of time to work on the new box designs, convene a math congress. Have a few students share the ways in which they used the price of one box to help with calculating the pricing of another. As each idea is put forth, examine it with the community and press for generalization.

As the math congress progresses, you will be introducing the open array to represent the partial products students are using and sharing, but be sure to connect the array directly to the context. Explain that you are drawing the box—do not call it an open array because it is not an open array yet. The lines placed on one rectangle to show its relationship to another smaller one represent only the ideas that students use as they explain the relationships they found—thus the lines are boxes within boxes. Stay within the context and use the students' graph paper blueprints to demonstrate the ideas expressed.

- Have a few students share how they used the price of one box to calculate the price of another and push for generalization of that strategy.
- Ask students to use the graph paper arrays to illustrate their strategies. Record their box designs as open arrays.

A Portion of the Math Congress

Toni (the teacher)**:** Chloe and Chris, would you begin our congress by sharing the idea you used? You started off counting every square, didn't you? And then you had an interesting idea that was a nice shortcut.

Chloe: We put one box on top of another. Then we didn't have to count anymore.

Chris: See, here's the 5 by 5 box. We knew that would sell for $25. Then we made a 4 by 5 and we put it on top. See, it's just 5 truffles less. We used the 5 by 5 for the 5 by 6 box, too. That one was 5 more…so the 5 by 6 would be $30. *(Uses graph paper arrays and puts one on top of the other.)*

Toni: Let me make a picture of what you are saying. Here's your 5 by 5. *(Draws a square.)* To figure out the 5 by 6, you added 5? *(Adds a 5 × 1 array onto the right side of the square, effectively turning the 5 × 5 into a 5 × 6.)*

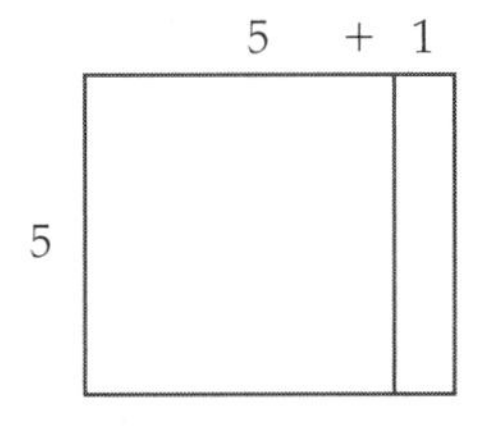

(Finishes the drawing, writing 25 in the area of the 5 × 5 and 5 in the area of the 5 × 1, and then writes an equation to represent the array.)

$5 \times 5 + 5 \times 1 = 5 \times 6$

I'll put parentheses around the parts that are the boxes to prevent any confusion. *(Adds parentheses:)*

$(5 \times 5) + (5 \times 1) = 5 \times 6$

Is this a picture of what you did?

Chris: Yep. Exactly.

Toni: Ok. Let me make the other one, too.

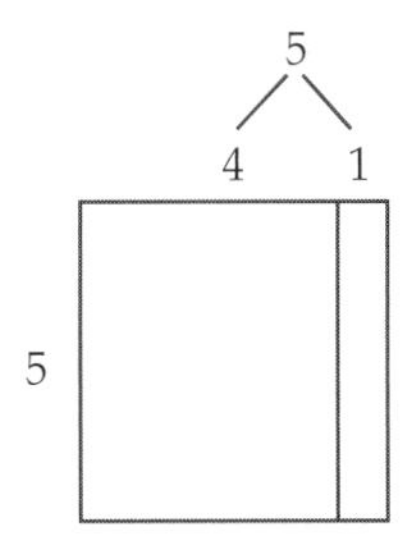

continued on next page

Author's Notes

Toni uses students' explorations and ideas as a jumping-off point to introduce the open array and formal notation. Presented within the context of boxes, the array and the notation are not abstractions, but directly linked to and representative of the realistic context that the students have been investigating. Over time, the open array will take the place of the boxes, becoming a tool for Toni to use as she represents student computation strategies during minilessons. Much later, the open array will become a tool to think with.

continued from previous page

Chloe: *(Later, after several other boxes and cases of using the distributive property have been discussed.)* I'm noticing something that I didn't see before. Look at the numbers. *(Points to the board where $(5 \times 5) + (3 \times 5) = 8 \times 5$ is written.)* The numbers have to add up. See, 5 + 3 equals 8.

As the congress continues, Toni represents both the open array and an equation for several boxes. After several equations have been listed for a variety of boxes, students begin to notice patterns. It is important in this community of mathematicians to look for patterns and wonder about them.

Toni: Who can put in your own words what Chloe means? Turn to the person next to you and talk about this.

Giles: I think what she means is that you can use smaller boxes to make bigger boxes. You just have to pick ones that fit. To make a fit, the numbers have to add up.

Paraphrasing and pair talk are powerful pedagogical strategies for heightening reflection on an important idea.

Toni: So if I wanted to know about 7 by 5, what small boxes could I use?

Giles: You could use 5 by 5 and 2 by 5.

Toni is now on her way toward helping her community generalize the distributive property.

Chloe: Or 3 by 5 and 4 by 5.

Sam: Or 7 by 2 and 7 by 3. Lots of ways. Maybe you could use three or four boxes, too. Not just two. You just to have to make sure the boxes fit.

Reflections on the Day

Today students used the truffles context to create different blueprints for boxes. These blueprints or arrays became a critical tool for students to explore the distributive, associative, and commutative properties of multiplication, as well as big ideas like equivalence and congruence. The math congress was used as a vehicle to share students' explorations and highlight some of the big ideas they were developing. One key component of this congress was the emergence of the open array as a tool to represent and explore partial products—a visual demonstration of the distributive property of multiplication. The array was considered to be a picture of a box of truffles arranged in rows and columns. At this point, the array is still directly connected to the context—it is a rectangular blueprint of a box design. As the unit progresses, the open array will emerge as a model to represent computation, and then as a tool to think with.

DAY FIVE

Blueprints for Muffles' Assortment Boxes

Materials Needed

Quick images (Appendix C)

Before class, prepare transparencies of the quick images on page 33.

Overhead projector

Muffles' new boxes poster (Appendix D)

2 x 5 box blueprints (Appendix E)—at least ten blueprints per pair of students.

Before class, cut out eight of the 2 × 5 box blueprints.

Student recording sheet for the new boxes investigation (Appendix F)—one per pair of students

Chart-size graph paper (with one-inch squares)—a few sheets per pair of students

Scissors—one pair per pair of students

Markers

Today's math workshop begins with a minilesson using a string of related quick images. The open array is used to record student strategies. A new investigation about blueprints for Muffles' assortments is then introduced. Students explore what sizes and configurations of large one-layer boxes for Muffles' assortments can be built using the small 2 × 5 boxes (the template for Muffles' original boxes). Students also draw blueprints on graph paper showing the small arrays that they arrange together to make larger boxes. This context lays the terrain for the work of the next three days as you continue with the development of the open array model. The investigation continues to support the use of partial products, and open arrays as representations of them, but still in the context of blueprints.

Day Five Outline

Minilesson: A Multiplication String

- Work on a string of quick images designed to encourage students to use partial products to find the products of larger arrays.
- Use the open array to record students' strategies, delineating the smaller arrays used.

Developing the Context

- Introduce a new truffles box scenario and explain that students will need to use 2 × 5 box blueprints to create new designs for larger truffles boxes.

Supporting the Investigation

- Note students' strategies as they work on new box designs.

Minilesson: A Multiplication String (10–15 minutes)

This mental-math minilesson uses a string of quick images (Appendix C) designed to encourage students to use partial products to find the products of larger arrays. Show one image at a time briefly on an overhead projector and then turn off the light on the projector or cover the image as you ask students to determine how many truffles could fit into that array, and to share how they know. Record the strategy shared on an open array, delineating the smaller arrays used. For example, if a student says for the third problem, "I saw two 2×5 boxes. I knew that was 10 and 10, so the answer is 20," use the quick image shown to highlight the strategy. Be sure to indicate (or ask) where the "10" is and what the dimensions of the new array would be (how $2 \times 5 + 2 \times 5 = 4 \times 5$).

- Work on a string of quick images designed to encourage students to use partial products to find the products of larger arrays.
- Use the open array to record students' strategies, delineating the smaller arrays used.

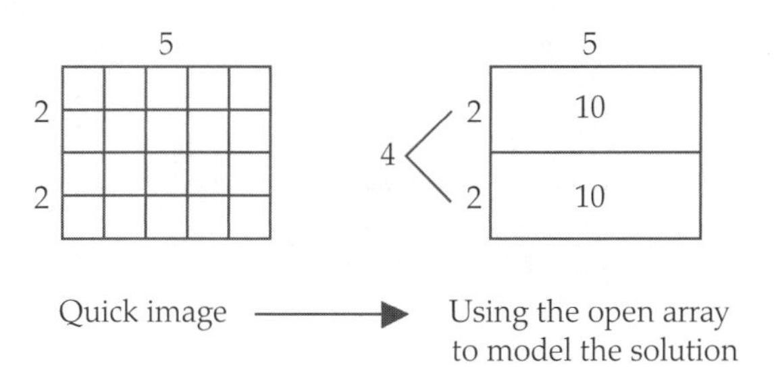

String of related quick images:

2×5

1×5

4×5
(made with two 2×5 boxes)

5×5
(made with two 2×5 boxes and one 1×5 box)

2×10
(made with two 2×5 boxes)

4×10
(made with four 2×5 boxes)

Behind the Numbers

The numbers in the string were chosen to develop the idea that small arrays can be used to build larger arrays. The arrays have been kept small and only two are used, a 2×5 and a 1×5, to allow students to determine what is shown since the arrays are seen only briefly. The quick image technique encourages students to move away from counting by ones as a way of figuring out the dimensions of the array, and the number of square units resulting from multiplying those dimensions where truffles can be placed, and, instead, to consider other strategies, such as skip-counting and using partial products. The string also provides an opportunity to use the open array as a representation.

A Portion of the Minilesson

Toni (the teacher)**:** I'm going to flash the image quickly. At the count of 3: 1, 2, 3. *(Flashes the 2 × 10 image.)* OK, thumbs-up when you're ready to share. Tanisha?

Tanisha: It was 20 truffles altogether.

Toni: And how did you know that?

continued on next page

Author's Notes

Toni prepares the students for the quick image by counting to three. This ensures that everyone is ready.

Toni supports communication by asking the student to explain her thinking.

continued from previous page

Tanisha: I saw two 2 by 5s, but they were put together differently than before. The other array—the 4 by 5—was almost a big square, but this one was different: it was a long rectangle.

Toni: It might help to think about the dimensions of this long rectangle.

Toni shows respect for the student's ideas by using her language ("a long rectangle"), but still pushes her to be more explicit.

Tanisha: I think it might be a 2 by 10. The short side was 2 and the long side was 10 because 5 and 5 makes 10.

Toni: So here's the quick image I showed you. *(Puts the image on the overhead and leaves it up to record on:)*

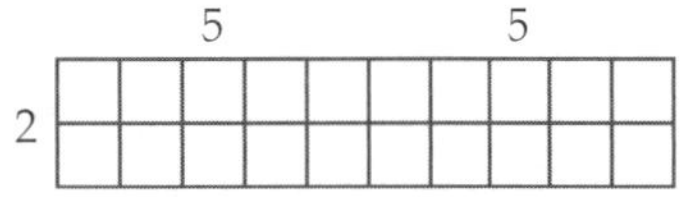

5 5

2

So two 2 by 5s is 2 by 5 plus 2 by 5. Someone help me understand where the 2 by 10 is that Tanisha is talking about. And what's the 20?

Toni uses the open array as a tool to communicate the student's strategy. She also checks for understanding in the other students by asking them what the dimensions of the new array would be.

Meta: The 20 is the inside, the number of squares inside where the truffles go.

Toni: How do you know that?

Meta: Put the two smaller arrays together. A 2 by 5 is 10. So mark 10 in each of those smaller arrays. Those smaller arrays make up the bigger array.

The open array is used as a tool to model partial products. The formal notation is not merely an abstraction, but is directly connected to the mathematical ideas students are exploring.

Toni: *(Returns to the model previously drawn and models Meta's ideas:)*

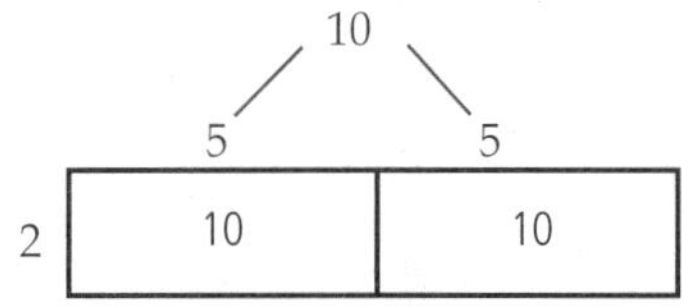

10

5 5

2 10 10

$(2 \times 5) + (2 \times 5) = 2 \times 10 = 10 + 10 = 20$

Assessment Tips

It is sometimes helpful to have students record their mental-math strategies in math notebooks. A benefit is that you have a record of how each student's thinking evolves over time. You can then make copies of student notes and add your own notes about their thinking. If you do not use math notebooks during your minilessons, use sticky notes or index cards to jot down notes about students as you see them use interesting strategies and develop an understanding of the array. Use one note or card for each student. At the end of the day, you can make a copy of the string, attach the sticky note or index card to it, and place it in the student's portfolio.

Developing the Context

Display the Muffles' New Boxes poster (or Appendix D) and continue the Muffles story as follows:

> *One day, as he is packing truffles, Muffles' assistant, Patricio, has a great idea. He tells Muffles, "You know, you can put boxes together and make a new, larger box for your assortments. Look, if I take two boxes and put them together, I have a new box!" You could put the vanilla truffles in one box and the dark chocolate truffles in the other. You could sell this two-flavor assortment box for $20." Muffles is fascinated by this idea and wonders, "What sizes of larger boxes can be made using several of the original boxes, put together?"*

Introduce a new truffles box scenario and explain that students will need to use 2 x 5 box blueprints to create new designs for larger truffles boxes.

Tell students that they will use 2 × 5 box blueprints to explore the following questions:

- *What sizes and shapes of boxes can be made from several of the original boxes put together?*
- *How many flavors would each larger assortment box hold if Muffles always fills a 2 x 5 section of the box with one type of truffle?*

Facilitate a brief group discussion on these questions using cutouts of the 2 × 5 box blueprints in Appendix E. Clarify that Muffles will only use rectangular boxes. Lay out three different arrangements of boxes, A, B, and C, and explain that Muffles only wants boxes that are rectangular, of type A and C, and with only one layer.

Figure A

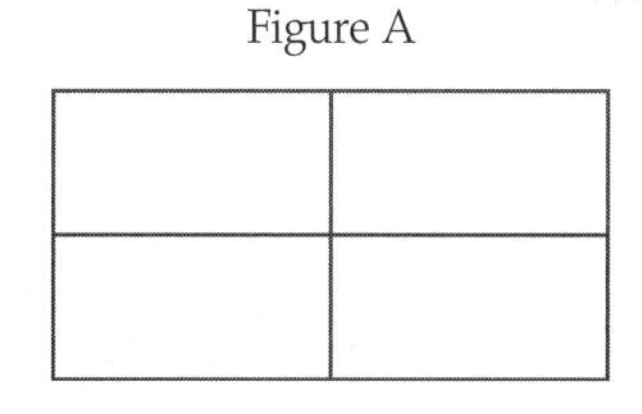

Figure B

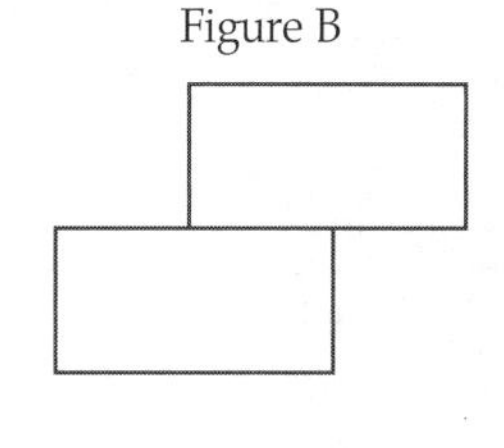

Figure C

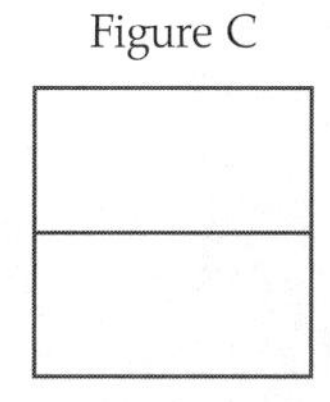

As students share their ideas, model what they are saying with the 2 × 5 blueprints and ask how they would label these new arrays. After this brief discussion, provide students with multiple copies of the 2 × 5 blueprints (Appendix E) and send them off to investigate.

Supporting the Investigation

The purpose of this investigation is not to find combinations for different types of truffles (strawberry and vanilla, or chocolate and strawberry) but to determine how many different larger one-layer boxes can be designed out of the 2 × 5 small boxes. As students build new arrays, you will see them exploring a number of big ideas. Look for these big ideas as you confer with students:

Note students' strategies as they work on new box designs.

✦ As one dimension doubles, the other halves. Generalized, this idea can be explained with the associative property: 2 × (2 × 5) = (2 × 2) × 5.

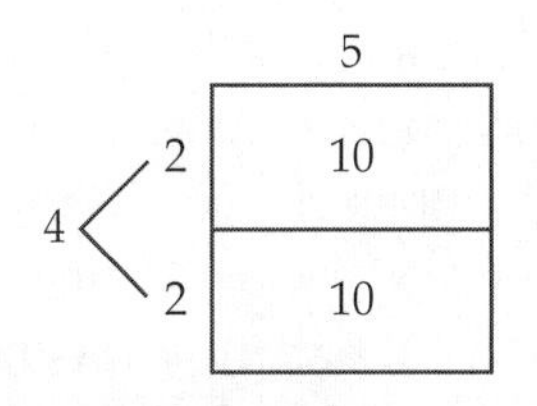

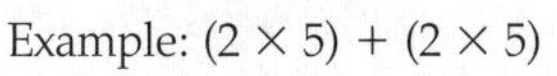

Example: (2 × 5) + (2 × 5)

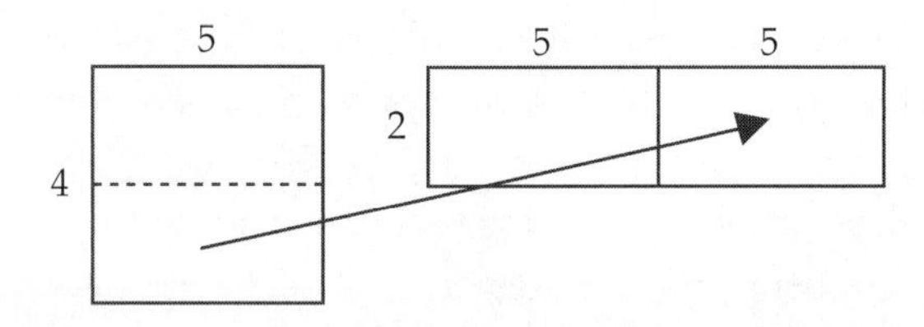

Example: 4 × 5 = (2 × 2) × 5 = 2 × (2 × 5)

- ✦ Factors and multiples are related; all the products—the numbers of truffles the various boxes will hold—are multiples of ten since the original array was 2 × 5.
- ✦ Some boxes are congruent and can be disregarded when determining the total number of different boxes that are possible.
- ✦ Since the original box was 2 × 5, the dimension of one side of the box is always an even number and the dimension of the other side will be a multiple of five.

Appendix F Student recording sheet for the new boxes investigation

Number of 2 × 5 Arrays Used	Total Number of Squares	New Blueprints Built
2	20	2 × 10 4 × 5

Once students have had sufficient time to explore new box designs, hand out recording sheets (Appendix F). As students work, walk around and take note of the strategies and big ideas you see being constructed. Confer with students as needed to support and challenge their investigation. Keep track of which big ideas students are exploring, and of how you might use their explorations in a later math congress to scaffold the learning of the community. Because students may be building new arrays without actually thinking about the resulting new dimensions, be sure to ask them how they would label their new blueprints for larger boxes. Have them draw blueprints on large graph paper, showing the arrangements of the smaller 2 × 5 boxes. The drawing of the blueprints is an important part of the progression in developing the open array model.

To help students think about the big ideas that come up during their explorations, ask them to cut out their new box designs and draw the smaller 2 × 5 arrays on them. Then ask them to use the 2 × 5 blueprints to model *how* the new, larger arrays were constructed. For example, if they cut out a 2 × 15, you might ask,"How is this 2 × 15 array related to the original 2 × 5 you used to build it?"You might also challenge students to think about the equivalence that is occurring by juxtaposing two arrays and noting that both arrays hold the same number of truffles. For example, the question"How is a 2 × 15 related to a 6 × 5?" may help students construct the associative property. Here the role of the 2 × 5 array in relationship to the newly cut-out arrays is critical. You might also—depending on how well a pair of students understands what is occurring—introduce the mathematical notation for this idea: $2 \times 15 = 2 \times (3 \times 5) = (2 \times 3) \times 5 = 6 \times 5$.

As students build and label arrays and fill in their recording sheets, some may notice patterns. For the arrays where 2 is a constant dimension, the other dimension increases by 5: 2 × 5, 2 × 10, 2 × 15, etc. For arrays where 5 is a constant dimension, the other dimension increases by 2. Ask *why* this pattern is happening. An important big idea for students to develop is that the patterns they are noticing are directly related to how many 2 × 5 arrays they are using and how they are arranging these 2 × 5 arrays.

Differentiating Instruction

Many students will be challenged just by building a variety of boxes and representing them as arrays with dimensions. Other students may need a greater challenge so encourage them to find all the possibilities and to prove they have them all. Since Muffles makes truffles in only ten flavors, the largest number of 2 × 5 boxes he can use is ten (for an assortment that includes one 2 × 5 box of each flavor). The smallest number of boxes he can use is two (for an assortment containing two flavors; one 2 × 5 box would not be an assortment). There are twenty unique boxes that can be made with at least two boxes and at most ten boxes. The following arrays are possible: 2×10; 2×15; 2×20; 2×25; 2×30; 2×35; 2×40; 2×45; 2×50; 4×5; 4×10; 4×15; 4×20; 4×25; 6×5; 6×10; 6×15; 8×5; 8×10; and 10×10.

Inside One Classroom

Conferring with Students at Work

Toni (the teacher)**:** *(Spends some time observing the students at work before she speaks.)* Help me understand what you're doing here.

Justin: We noticed a pattern. Look, if you build the blueprint going this way, the length goes up by 5.

Ellen: You count, like 5, 10, 15, 20.

Justin: Yeah. And this side stays the same. It's always 2.

Toni: Wow! That's really interesting. I wonder why it's happening.

Ellen: Because you keep adding 5.

Toni: Adding 5. Could you show me on your blueprints where the plus 5 is? Here's the original 2 by 5.

Justin: Yeah, then it goes 2 by 10. *(Points to his recording sheet.)* Then 2 by 15, 2 by 20, 2 by 25. We think it would just keep going like that.

Toni: And Ellen said you were adding 5. Help me understand where that 5 is.

Ellen: See, 15 and 5, 20; 20 and 5, 25. It keeps getting longer by 5.

Toni: *(Points to their recording sheet.)* That's an interesting observation, but I'm still wondering what the 5 is, because I notice that the numbers you wrote here for the total number of squares in the array are 10, 20, 30, 40, and 50.

Justin: That's weird. We thought it was going up by 5.

Ellen: It is. Look, every time we built a new array, we got 5 more squares in each row for truffles.

Toni: Let's look at what's happening with the blueprints. Here's a 2 by 5. Then what happened to make this new array?

Justin: We added 5.

Ellen: No, wait, I think I get it. We're not adding 5, we're adding 2 groups of 5. That's where the 10 is coming from over here. *(Points to the "Total Number of Squares" column on their recording sheet.)*

Justin: So what is this that's going up by 5?

continued on next page

Author's Notes

Watching students work can help a teacher make sense of their strategies. These observations can then be used to decide what direction the interaction will take. Toni listens for awhile before she asks for clarification.

Toni recognizes that although the students have found a pattern, they may not know the mathematical reason for pattern.

Toni tries to create disequilibrium by referring to another pattern on the students' recording sheet—that of growth in box capacity. Her questioning causes them to reflect and to differentiate the two patterns.

Students often confuse a change in a linear dimension (the length and width) with a change in area (square units). Toni stays grounded in the context asking them to use their drawing to show the increasing five. Each row increases with five squares, the length increases by five inches, but the total area of square units increases by ten, the result of the addition of two rows of five squares.

continued from previous page

Ellen: I think the length of the box is getting longer by 5, but the amount of truffles it holds is getting bigger by 10. If we only added 5, it would look like this. *(Covers the bottom row of the 2 × 5.)*

One of the powers of models is that students can realize the solution to their struggles through modeling what is happening in the situation.

Toni: Looks like you're onto something here. A 2 by 10 is really a 2 by 5 plus another 2 by 5. *(Writes:)*

$2 \times 10 = (2 \times 5) + (2 \times 5)$

The parentheses help us think about the groups.

Toni uses formal notation in the context of representing the mathematical ideas students are using.

Justin: Oh, I get it. This is where the 10 is: it's the 2 fives that make this length—the truffles in this row.

Toni: So how would you show what is happening with some of these other arrays you've built?

Then she gives the students an opportunity to try this new notation.

Justin: *(Writes:)*

$2 \times 5 + 2 \times 5 + 2 \times 5 = 2 \times 15.$

Toni: I'm going to see what other groups are doing. Make sure you keep good notes about what you're doing here. I think what you've found might be helpful to other kids who have noticed this pattern, but aren't sure what is making it happen. Perhaps you could share your ideas in our math congress tomorrow.

Toni points out how their work is connected to that of other students. This subtle technique—which promotes the concept that ideas can be used as tools to help other students—is one way to build a mathematical community.

Assessment Tips

Reflective writing can be used as a way to assess individual growth and development. Consider asking students to reflect on their learning from today's investigation in light of Muffles' original question: *What sizes of one-layer boxes can he make by combining the 2 x 5 boxes?* Ask them to write a letter to Muffles in which they share their findings.

You might also find it helpful to make a copy of the graphic of the landscape of learning provided in the unit overview (page 11). As you continue with the unit, you can color in the big ideas and strategies you see students developing.

Reflections on the Day

Today students had an opportunity to construct different arrays using the original 2 × 5 box blueprint. Some students explored ways that arrays with the same total number of squares could have different dimensions, and they may have applied doubling and halving to the task of finding arrays that are equivalent in the number of sqaures, but have different dimensions. Other students may be working to determine if they have found all the possibilities. On Day Six, students will spend more time on this investigation and will make posters of their designs for assortment boxes.

DAY SIX

Blueprints for Muffles' Assortment Boxes

Today's math workshop begins with a minilesson similar to the minilesson on Day Five. It makes use of quick images again as a way to support students in progressing beyond counting. An open array is used to record student strategies. After the minilesson, students continue working on their designs for Muffles' assortment boxes and prepare posters for the math congress to be held on Day Seven.

Day Six Outline

Minilesson: A Multiplication String

- Work on a string of quick images designed to encourage students to use partial products to find the products of larger arrays.
- Use the open array to record students' strategies, delineating the smaller arrays used.

Preparing for the Math Congress

- Give students time to complete their new box designs and then ask them to prepare posters of their strategies.
- Encourage students to determine if they have found all the possible boxes.
- Conduct a gallery walk to give students a chance to examine each other's box designs and strategies.
- Plan to focus the congress (to be held on Day Seven) on developing the array as a tool for modeling the commutative, distributive, and associative properties.

Materials Needed

Quick images (Appendix C)

Before class, prepare transparencies of the quick images listed in the string on page 40.

Overhead projector and overhead markers

Students' work from Day Five

Large chart paper—one sheet per pair of students

Sticky notes—one pad per student

Large chart pad and easel

Markers

Minilesson: A Multiplication String (10–15 minutes)

- Work on a string of quick images designed to encourage students to use partial products to find the products of larger arrays.
- Use the open array to record students' strategies, delineating the smaller arrays used.

This mental-math minilesson uses a string of related quick images (Appendix C) designed to encourage students to use partial products to find the products of large arrays. As you did on Day Five, show one image at a time briefly and then turn off the light on the projector or cover the image as you ask students to determine how many truffles are shown and to explain how they know. Record the strategy shared on an open array, delineating the smaller arrays used.

String of related quick images:

2×5

1×5

3×5
(made with one 2×5 box and one 1×5 box)

5×4
(made with two 2×5 boxes vertically)

4×5
(turn the 5×4 array 90 degrees)

5×5
(made with two 2×5 boxes and one 1×5 box)

Behind the Numbers

The numbers in the string were chosen to continue to develop the idea that small arrays can be used to build larger arrays. Only two small arrays are used, a 2×5 and a 1×5, to give students a better chance of determining what is shown since the arrays are seen only briefly. As you use the 2×5 and 1×5 to make the other images, leave only a slight space between them to enable students to see them easily while keeping them intact as an array. The quick image technique encourages students to progress beyond counting by ones in figuring out the area and to consider other strategies, such as skip-counting or using partial products. The string also provides an opportunity to use the open array as a representation. The first two problems are used to make the third. The fourth and fifth problems will generate a discussion on the commutative property. The last problem requires partial products again.

Preparing for the Math Congress

- Give students time to complete their new box designs and then ask them to prepare posters of their strategies.
- Encourage students to determine if they have found all the possible boxes.
- Conduct a gallery walk to give students a chance to examine each other's box designs and strategies.
- Plan to focus the congress (to be held on Day Seven) on developing the array as a tool for modeling the commutative, distributive, and associative properties.

After the minilesson, allow students time to finish their work on the assortment box blueprints, and then ask them to prepare posters for a math congress to be held on Day Seven. As students work, help them examine the question of whether or not some of their boxes are congruent. Remind them that the goal is to develop blueprints of *different* designs. Where you think it is appropriate, press students to determine if they have found all the possible boxes. This question may be just the thing that will bring students to consolidate and construct the big ideas this investigation is designed to support. Remind them that the task is to find different boxes, not different arrangements of the flavors, and encourage them to find a way to organize their work. For example, they might record all the different boxes for assortments of two flavors, (2×10, 4×5) and then for three flavors, etc., or they might organize by keeping one dimension constant (all the boxes with eight rows, 8×5, 8×10).

Plan on having a gallery walk toward the end of the math workshop today. Since some students will probably have designed boxes that others have not thought of, the gallery walk will allow students to examine a variety of boxes and strategies. If they see a box they did not make, they will be pushed to examine ways to systematize their work, and they will need to analyze for equivalency and congruence. Discussion of the commutative property may also arise during the gallery walk as students determine congruency (such as 8×10

and 10 × 8). Discussing strategies and different boxes during the gallery walk will allow you to focus on just a few big ideas during the congress on Day Seven.

For the gallery walk, pass out pads of sticky notes and suggest that students use them to record comments or questions. These notes can be placed directly on the posters. Display all the posters around the room and have the students spend about fifteen minutes walking around, reading and commenting on the mathematics on the posters. Then provide time for students to read the comments and questions attached to their own posters.

Tips for Structuring the Math Congress

Plan on focusing the congress on Day Seven on developing the array as a tool for modeling the commutative, distributive and associative properties of multiplication. Choose two or three posters to scaffold the conversation. For example:

- Begin with the work of students who noticed a pattern in their recording sheet (the pattern went by twos: 2 × 5, 4 × 5, 6 × 5, etc.) but do not yet understand why this pattern occurs.
- Then move to a pair of students who used the array to model distributivity: how a 4 × 5 can be thought of as two 2 × 5 arrays because (2 × 5) + (2 × 5) = 4 × 5.
- Then ask a pair of students who examined the other pattern (how the length of the array increases by five) to display their poster, and ask the class to consider, in light of the previous discussion, what might be making this pattern happen. Use notation to represent these ideas—(2 × 5) + (2 × 5) = 2 × 10—and to highlight which dimension stays the same and which dimension changes.
- If any students have developed a way to explain doubling and halving (a way to model why 4 × 5 = 2 × 10), you might discuss their poster next. A piece of work like this may be too difficult to consider at the beginning of the congress, (especially if students have not yet considered this idea). However, the prior discussion of distributivity will support students as they think about doubling and halving and the associative property.
- At the end of the congress, shift the conversation to the questions "Do you have all the boxes?" and "How can you be sure?" If some students have systematized their work, let them defend their systems.

Reflections on the Day

Quick images were used again today to support students in envisioning how small arrays could be used to make larger arrays. The minilesson also allowed you to use the open array to model the computation strategies students were using as they found solutions for the quick images. A gallery walk provided time for students to examine each other's work and to rethink whether they found all of the possible boxes and how they might have proceeded more systematically.

DAY SEVEN

Assortment Boxes

Materials Needed

Chart-size graph paper (with one-inch squares), as needed

Students' posters from Day Six

Large chart pad and easel

Markers

The day begins with a minilesson—a string of related problems designed to stimulate a discussion of the commutative property and the ten-times pattern, providing a nice connection to the work on place value in the first two days of the unit. The open array is now used to record students' strategies, as well. A math congress on the work of designing the assortment boxes is then held.

Day Seven Outline

Minilesson: A Multiplication String

- Work on a string of problems designed to encourage students to use partial products to make other arrays.
- Use the open array to model students' strategies, this time without delineating the smaller arrays.

Preparing for the Math Congress (continued from Day Six)

- Give students time to review the comments on their posters from the gallery walk on Day Six.

Facilitating the Math Congress

- Discuss posters that will highlight the distributive and associative properties.

Minilesson: A Multiplication String (10–15 minutes)

This mental-math minilesson builds on students' prior experiences with quick images—using partial products to make other arrays—but now only the open array is used to model students' strategies. If some students still need the support of graph paper arrays, have chart-size graph paper handy on an easel so you can draw the arrays on it when necessary, but continue to encourage students to envision the open array as representing rows and columns. Suggest that the rectangle you are drawing is the outline of one of Muffles' boxes.

- Work on a string of problems designed to encourage students to use partial products to make other arrays.
- Use the open array to model students' strategies, this time without delineating the smaller arrays.

Do one problem at a time, giving enough think time before you start discussion. Record student strategies to the right of the string. Keeping the previous problems, answers, and representations in the string visible to the students will help them think about how the problems are related. Invite students to discuss the relationships among the problems.

String of related problems:

2×5

4×5

4×10

10×4

10×6

6×10

10×12

10×18

11×18

The numbers in the string were chosen for two reasons. First, they provide a transition from the quick images of the last two days, when the 2×5 box was shown, to a mental arithmetic approach in which students now have to imagine arrays being built and the open array is used as a model for representing *their* images and strategies. Second, the ten-times strategy explored on Days One and Two can be revisited. As students start using the array model with large numbers, the ten-times strategy and the distributive property will become critical. For example, 6×18 is easily solved when students think of it as $6 \times 10 + 6 \times 8$. Students may remember from the work they did on Days One and Two that a number times 10 moves over a place value—because of the place value inherent in our number system. In this minilesson, they revisit that idea but they are also challenged to understand the commutativity involved. It is easy to see why 6×10 is 60: $10 + 10 + 10$, etc. But it is a bit more difficult to see why $6 + 6 + 6$, etc., results in 60, until students realize that $6 \times 10 = 10 \times 6$. Since there are 6 tens (as well as 10 sixes), the 6 moves over to the tens column.

Behind the Numbers

Expect students to quickly find the products for the first two problems in the string. Because these problems are relatively easy, don't spend a lot of time discussing them, but draw the open arrays to represent student solutions. The third problem will probably give you a chance to place arrays together if students say the solution to that problem is double the previous one. Make sure that you build onto the open arrays properly to make new arrays with the dimensions in the problem and/or to match what students say.

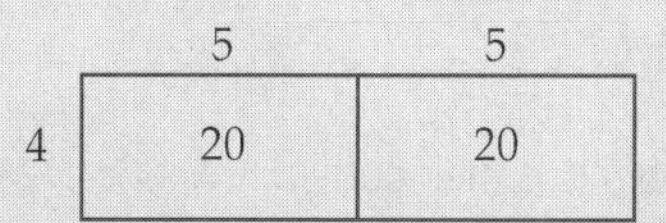

The third and fourth problems will bring up the commutative property, as will the fifth and sixth. Help students realize that, in essence, the array is rotated. Encourage them to think about how 120 truffles was just the right amount for 12 boxes, and remind them of the count-around on Day Three. The last three problems in the string require that students think through the solutions without the use of helper problems.

Preparing for the Math Congress (continued from Day Six)

Give students time to review the comments on their posters from the gallery walk on Day Six.

To revisit the mathematics of Day Six before you begin the math congress, suggest that math partners from Day Six take a few minutes to read the comments and queries about their posters. Ask them to prepare to come to the math congress to discuss a few of the ideas.

Facilitating the Math Congress

Discuss posters that will highlight the distributive and associative properties.

Gather the students in the meeting area. Have them sit next to their partners. Begin by inviting one pair of students to display and discuss their poster.

Inside One Classroom

A Portion of the Math Congress

Toni (the teacher)**:** So Alex and Mary noticed a pattern and left us with two questions: "Why does this pattern work?" and "Will it always work?" Justin and Ellen, I'm going to ask you to go next. *(Displays their poster next to Alex and Mary's.)*

Justin: We worked on the same thing and we noticed the same pattern—that the arrays were going by twos. Well, at first we thought it was adding 2, but then we realized it was really adding 10.

Toni: Let's have a show of hands. Who thinks you can put in your own words the ideas Justin is sharing with us? Quite a few of you. OK, Meg?

Meg: See, this is their pattern. *(Points it out on their poster.)* It's 2 by 5, 4 by 5, 6 by 5, 8 by 5. It looks like it's going up by 2 because you go 2, 4, 6, 8. But this *(points to another column on their poster, the price or total of the truffles)* is going up by 10. See, it's 10, 20, 30, and 40.

Ellen: That's why we put the chart here to help us explain, because at first we thought like Alex and Mary that the pattern was about adding 2. But you're not just adding 2, like 1, 2; you're adding 2 groups of 5. That's why it goes up by 10.

Toni: Comments or questions?

Pete: I just don't get it.

Toni: Try to think about what it is you don't get. Try to be more specific. Perhaps find a way to question. Just saying "I don't get it" doesn't help us understand what's puzzling you.

Pete: I think it is adding 2. Otherwise, why would it go 2, 4, 6, 8?

continued on next page

Author's Notes

Before moving on, Toni emphasizes the two questions posed by Alex and Mary. These are powerful questions for students to incorporate into their repertoire. Finding that there may not always be an answer to a problem or that the solution may not be found immediately is an important experience for students of mathematics to have. It develops a habit of mind that is based on resilience and persistence. Students need to understand that the questions they pose to themselves and the community are a big part of what it means to do mathematics.

To ensure that everyone has a way to enter this conversation, Toni checks to see if students understand by asking whether they can put Justin's idea into their own words. Many teachers ask: Does everyone get it? Students will often nod even if they don't. By asking who can explain, Toni ensures that only those who can will raise their hands, allowing her to see who needs further help.

Toni recognizes that the conversation may be moving too quickly for some students, and slows it down by asking for comments or questions.

Toni pushes students to think about what it is they don't understand. Although not understanding a solution is acceptable, saying "I don't get it" as a response is not. Asking Pete to be more specific forces him to think about his own thinking (metacognition). It also supports the dialogue that occurs next.

continued from previous page

Ellen: Do you see this? It's going up by 10.

Pete: So why can't you have two patterns? One goes up by 2 and the other goes up by 10.

Justin: Could we use these? *(Shows the arrays they have cut out.)* This might help a little. It helped us. Here's a box that's 8 by 5. But we made it with these 2 by 5 arrays. *(The conversation focuses on how students used arrays to explore what is creating the pattern. If one just "adds 2" one no longer has an array. The conversation then shifts to building arrays from other arrays.)*

The array becomes a critical tool to model what is happening. Many students confuse area and dimension. It is important to examine this difference.

Toni: How is this related to the 8 by 5 array? Pete, what do you think?

One way to support development is to return to students who have struggled with certain concepts to see if the new ideas that have been presented can help them resolve their questions.

Pete: It has four 2 by 5s in it.

Toni: So we can think of this array like this. *(Writes:)*

$8 \times 5 = 4 \times (2 \times 5)$

Toni records Pete's observation with the formal notation. This highlights the way that mathematical symbols are used as a form of communication.

Pete: Does that mean a 6 by 5 is a 2 by 5 plus a 2 by 5 plus another 2 by 5?

Justin: Yep. Three 2 by 5s. Hey, I just noticed that 3 times 2 is 6. So that's where the 6 is coming from!

Toni: *(Writes:)*

$6 \times 5 = 3 \times (2 \times 5) = (3 \times 2) \times 5$

Justin: Yeah.

Pete: You could also think about the 6×5 as a 4×5 plus a 2×5.

Toni: *(Writes:)*

$6 \times 5 = (4 \times 5) + (2 \times 5)$

So what happens when we go from a 6 by 5 to a 10 by 5? How would we express this change? I'd like you to turn and talk to your partner about this question: How does a 6 by 5 grow into 10 by 5?

Because the distributive property is such an important idea, Toni recognizes that some students may need time to continue to think about what is happening. She poses a related problem and uses pair talk to foster development. Using pair talk also helps her. As she listens to different groups, she makes mental notes about students' thinking and which students might need more support in the future.

Reflections on the Day

In the minilesson, the open array was introduced as a representation of student thinking, and the commutative property as well as the use of ten-times were revisited. Then students engaged in a math congress that was scaffolded to focus on the distributive and associative properties.

DAY EIGHT

The Big Mix-Up

Materials Needed

Big mix-up poster
[If you do not have the full-color poster (available from Heinemann), you can use the smaller black-and-white version in Appendix G.]

Box outlines (Appendix H)
Prior to class make one set of four box outlines per pair of students

Measurement tools (Appendix I)—one set per pair of students

Glue sticks—one per pair of students

Large chart paper—one sheet per pair of students

Chart-size graph paper (with one-inch squares), as needed

Large chart pad and easel

Markers

Math workshop begins with a minilesson, which today will be a string of related problems to encourage the use of the distributive property. Strategies are represented with the open array. After the minilesson, a new investigation is introduced that gives an intriguing twist to the Muffles story. In this investigation, students think about how to label wrapped boxes without actually seeing the blueprints that give their dimensions.

Day Eight Outline

Minilesson: A Multiplication String

- Work on a string of problems designed to encourage students to use partial products to solve more difficult problems.
- Use the open array to record students' strategies.

Developing the Context

- Explain to students that they will now investigate how to label wrapped boxes without being given the dimensions of the boxes.

Supporting the Investigation

- Take note of students' use of the distributive property and the kind of equations they use to represent their thinking.
- Ask students to glue their box outlines onto larger chart paper to share during the math congress.

Minilesson: A Multiplication String (10–15 minutes)

This mental-math minilesson builds on students' previous experiences with the distributive property. The string uses related problems designed to encourage students to think about what they know and to use partial products for solving more difficult multiplication problems. Do one problem at a time, giving enough think time before you start discussion. Record student strategies to the side of the string with open arrays, but have graph paper available so you can draw the arrays on it if necessary. Keeping the previous problems, answers, and representations in the string visible to students will help them think about how the problems are related. Invite students to discuss the relationships among the problems.

- Work on a string of problems designed to encourage students to use partial products to solve more difficult problems.
- Use the open array to record students' strategies.

String of related problems:

10×5

4×5

14×5

14×10

14×9

14×19

12×19

Behind the Numbers

The numbers in the string were chosen to continue to develop the idea that several small arrays can be used to build larger arrays. The focus is now on double-digit multiplication; the array is used both to model student strategies and to help students think about distributivity. Expect students to quickly find the products of the first two problems in the string. Because these two solutions are relatively easy to find, don't spend a lot of time discussing them, but draw the arrays to represent students' solutions. The third problem gives you a chance to add the two arrays together if students offer that strategy.

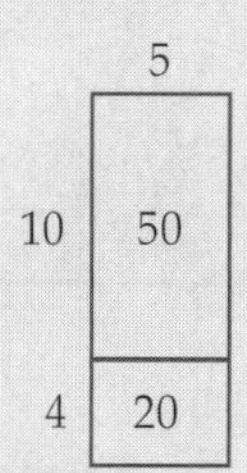

The next three problems are related in a similar fashion. The fourth problem in the string (14×10) doubles the product of the previous problem (because the factor of 5 becomes 10). Once again the distributive property underlies the partial products: $(14 \times 5) + (14 \times 5) = 14 \times 10$. Now another 14×5 can be attached to the right of the earlier drawn 14×5 array, forming a 14×10. The last problem in the string requires students to make their own helper problems.

Developing the Context

A new investigation is introduced with an intriguing twist to the Muffles story. In this investigation, students think about how to label wrapped boxes without actually seeing the blueprints that give the boxes' dimensions. The only tools they use are the arrays in Appendix I. With these open arrays as tools, students find ways to label box outlines. To solve the problem presented

- Explain to students that they will now investigate how to label wrapped boxes without being given the dimensions of the boxes.

Appendix G

The big mix-up poster

Appendix H

Box outlines

10 × 10	12 × 12
10 × 12	9 × 6
8 × 7	7 × 7
9 × 7	8 × 8
6 × 7	9 × 9
5 × 6	4 × 4
6 × 6	9 × 8

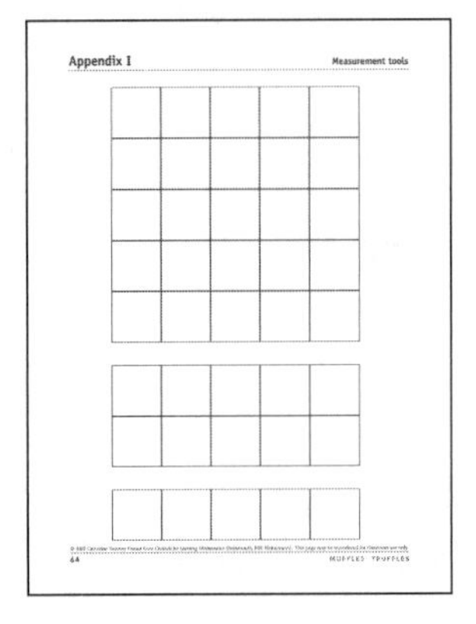
Appendix I

Measurement tools

in this investigation, students will need to apply some of the mathematical ideas developed thus far in the unit. Building up arrays to make larger ones is much easier developmentally than covering a given array with another in order to compare. Students are being asked to use the covering strategy here.

Display the big mix-up poster (or Appendix G), and tell the following story:

> *There is a big mix-up at the truffles shop! Muffles had decided to order some gold wrapping paper for his newly designed boxes. He asked Patricio to put the Muffles' Truffles logo on the top of each wrapped box and tie a bow around it. But now Muffles has a big problem. Patricio wrapped the boxes, but forgot to label them! Unwrapping the boxes would spoil the gold paper, but Muffles needs to know how many truffles are inside each box in order to know what to charge his customers. When Muffles looks at the boxes, he wonders, "How can I possibly figure out how many truffles are in each box?"*

Explain that you have outlines of the boxes and copies of a few of the previous blueprints that students might use to solve Muffles' dilemma. Assign math partners and pass out one set of box outlines (see Appendix H) and one set of measurement tools (Appendix I) to each pair of students. Do *not* give students multiple copies of the small arrays in Appendix I. Only one copy of each of the small arrays (measurement tools) is provided intentionally to encourage and support students to *envision* filling the open array. Providing more than one copy allows students to simply cover the whole array and count the number of square units, which offers little potential for learning.

Facilitate a brief preliminary discussion in the meeting area before the students set to work, to ensure that they understand the investigation and to show them how to record. Have them outline the small arrays they use directly onto the box outlines (open arrays) and label them.

Supporting the Investigation

As students work, walk around and take note of the strategies you see. Confer with students as needed to support and challenge their investigation. Here are some strategies you might see for determining the dimensions of the boxes:

- Take note of students' use of the distributive property and the kind of equations they use to represent their thinking.
- Ask students to glue their box outlines onto larger chart paper to share during the math congress.

✦ Using the 2 × 5 array to figure out dimensions of other arrays by
  - aligning one side of the 2 × 5 array with that of another open array, or
  - laying the 2 × 5 array on top of an open array to determine its dimensions.

  As you confer with these students, encourage them to consider what the number of squares would be of the arrays they are imagining.

- Separating out arrays where one dimension is the same (e.g., finding all the arrays that have two as a dimension).
- Using the square (5×5) as a template for finding other arrays. When you confer with these students, encourage them to imagine increasing squares—how a 5×5 can become a 6×6 by adding two 1×5 arrays plus one unit square. *[See Figure 4]*
- Using the distributive property in a number of different ways: for example, seeing 12×10 as 10×10 (perhaps envisioned as four 5×5 arrays or ten 5×2 arrays) and then imagining the piece left as two 2×5 arrays.

Figure 4

As you confer with the students at work, help them think about the mathematical ideas that underlie their strategies. As you observe students working with or developing critical big ideas, be sure to ask them to label their arrays. The box outlines with their labels of the small arrays should be glued onto larger chart paper in order to label the dimensions. These posters can be the basis of your math congress on Day Nine.

Assessment Tips

This investigation allows you to observe how students are using some of the mathematical ideas developed over the past seven days. Take careful notes on how students use and seem to understand the distributive property, and on the kinds of equations they write to represent the ideas they are using. Also be mindful of students who are confused by the idea of an open array and have difficulty visualizing the meaning of 5×8 when there are no squares to count. They may need more time to work with arrays. They may also need to explore multiplication further. One way to do this is to work with small groups of eight to ten students on guided minilessons, crafting the strings of problems to meet the students' needs.

Reflections on the Day

In the minilesson today, student strategies were modeled with the open array. Because this mathematical model has been developed over the course of this unit, you can now use the open array as a tool to represent student strategies in your multiplication minilessons.

After the minilesson, students returned to the context to figure out the dimensions of a set of box outlines. As they worked, students needed to think about and use some of the mathematical big ideas (like the distributive property) previously developed in this unit. Students also began creating posters for the math congress to be held on Day Nine.

DAY NINE

The Big Mix-Up

Materials Needed

Students' posters from Day Eight

Large chart paper, as needed

Sticky notes—one pad per student

Large chart pad and easel

Markers

Math workshop begins with a string of related problems designed to continue supporting the use of the distributive property for multiplication over addition *and* over subtraction. The open array is used to record strategies. Students finish making their posters on the problems in the big mix-up story context, and then have a gallery walk followed by a math congress.

Day Nine Outline

Minilesson: A Multiplication String

- Work on a string of problems designed to encourage students to use both partial products and ten-times.
- Use the open array to record students' strategies.

Preparing for the Math Congress

- Give students time to finish their posters and then conduct a gallery walk to give students a chance to review each other's work.
- Plan to focus the congress on using the open array to solve the problems.

Facilitating the Math Congress

- Facilitate a discussion to support students' understanding of how small arrays can be used to determine the dimensions of larger arrays.

Minilesson: A Multiplication String (10–15 minutes)

This mental-math minilesson is designed to encourage students to use both partial products and landmarks in the number system, such as the use of ten-times. Do one problem at a time and give enough think time before you start discussion. Record student strategies to the side of the string. Invite students to think about how the open array is being used to model specific mathematical ideas.

- Work on a string of problems designed to encourage students to use both partial products and ten-times.
- Use the open array to record students' strategies.

String of related problems:

10×10

12×10

12×9

12×19

14×9

14×21

Behind the Numbers

The numbers in the string were carefully chosen to develop the use of partial products and the use of ten-times. The first problem is the scaffold for the string. The second problem is two more groups of ten and the third problem is one fewer. The fourth, fifth, and sixth problems require students to consider ways to make them friendlier—to find partial products on their own. Accept a variety of solutions and model them on the open array.

Preparing for the Math Congress

Give students time to finish their posters from Day Eight. Display the posters around the meeting area. Then explain that the class will have a gallery walk to look at the posters before you start a congress. Pass out small pads of sticky notes and suggest that students use them to record comments or questions. These notes can be placed directly on the posters. Allow about fifteen minutes for everyone to walk around, reading and commenting on the mathematics on the posters. As students walk around, think about how you will structure the congress. Give students a few minutes to read the comments and questions posted on their own work and then convene a math congress.

- Give students time to finish their posters and then conduct a gallery walk to give students a chance to review each other's work.
- Plan to focus the congress on using the open array to solve the problems.

Tips for Structuring the Math Congress

You will want to focus your math congress on using the open array to solve the problems. Look for samples of work that can be used to model the big ideas (like the distributive property) or strategies (like doubling or tripling) that you want students to consider in relation to the open array.

Facilitating the Math Congress

Bring children to the meeting area. Have two or three pairs of students share and facilitate the discussion.

- Facilitate a discussion to support students' understanding of how small arrays can be used to determine the dimensions of larger arrays.

A Portion of the Math Congress

Author's Notes

(Students have been sharing strategies for labeling their open arrays, but some students have become confused. The array under question is an 8 × 7.)

Nellie: I'm not really sure how to label this little corner. *(Points to the missing 3 × 2 piece.)* We used a 5 by 5 and we know there is a 2 by 5 on this side *(to the right)* and a 3 by 5 here *(on the top)*. But then we had this little hole.

Toni (the teacher)**:** Does anyone have a way to explain how to label this small array?

Recognizing that there may be some who are confused or unsure of the ideas presented thus far, Toni spends time exploring this further.

Kurt: I think it is 3 by 2.

Alex: No way.

Toni: Those are not convincing arguments, Alex and Kurt. Is there a way to convince us that this is not a 3 by 2, or that it is?

Alex: Sure. How many truffles are in a 3 by 2?

Class: Six.

Toni: We could, of course, count to see if there are 6, but the squares aren't here for us to do that. Let's think about what we know so far. What's the size of this array on top?

Toni encourages the students to determine the label and provide a convincing argument. A convincing argument needs supporting evidence.

Alex: It's 3 by 5.

Toni: So how many rows would that be in this little one? Alex?

Alex: Three.

Toni: And how many columns will there be because of this 5 by 2 here?

To help students think about the distributive property, Toni brings the conversation back to the dimensions of the array.

Alex: Two. Oh, yeah, it's a 3 by 2.

Toni: So it seems this little hole was made by the size of those other two arrays. That's interesting. Now that we are sure of what the pieces are, what shall we label this box?

Alex: It must be 8 by 7.

Reflections on the Day

Today students continued to work with the open array. The mental-math minilesson focused on the use of the distributive property over addition and over subtraction as students examined representations of their strategies on open arrays. The math congress supported consolidation of understanding the ways in which small arrays can be used to determine the dimensions of larger arrays.

DAY TEN

A Day for Reflecting

Today's math workshop is focused on reflection as students create a wall display to represent the learning that has occurred over the course of this unit. To create this community display, students revisit the work they completed over the past nine days, retracing their thinking and charting the big ideas, strategies, and models that they examined in their investigations. Questions that were raised but left unexplored are posted as possible future explorations for the mathematical community.

Day Ten Outline

Building the Wall Display

- Work with students to create a wall display to highlight the work they have done throughout the unit.

Materials Needed

Large butcher paper for the wall display (eight feet long or more, depending on available space)

Students' work from throughout the unit

Drawing paper—a few sheets per student

Scissors and tape

Markers

Building the Wall Display

Work with students to create a wall display to highlight the work they have done throughout the unit.

Gather the students in the meeting area. Have them sit next to their math partners. Announce that you are going to create a wall display of all the work they have done over the past nine days as a record of their mathematical explorations and ideas. The display will have two purposes. First, it will allow students to revisit all their work and think about what they have learned and what questions they may still have. These questions might become the basis of future explorations. Second, because the wall display will be a place for reflection, the rest of the school community can also think about these ideas. Emphasize to the students that this is a document of *their* learning. And because it is a public document, it needs to clearly communicate the scope of their learning.

As part of creating this display, ask students to revisit the work they completed over the past nine days. To support them in this task, make a list of the various questions they explored. Post the first question and ask students to examine their work and think about what important ideas, strategies, and further questions came up in this discussion. As a classroom community, pick several pieces of student work that are representative of these big ideas, strategies, and questions. You may find that some questions have not yet been explored; these can be listed under a heading like "Our Questions for Future Exploration."

The primary purpose of a wall display is to give students an opportunity to reflect on their own thinking and then to consider how their ideas fit into and contribute to the mathematical learning of the entire classroom community. In this sense, it is a sociohistorical record. The wall display is also a way to emphasize the role that communication plays in mathematical life. It is one thing for students to communicate their ideas to their math partners and then to their classmates. It is quite another to communicate these ideas to a larger community. Doing so means sifting through their work and thinking about what is important and how they want to express these ideas. Students cannot do this without self-reflection, which is one of the critical tools they need to develop as learners.

The main goal here is to create a living document that accurately reflects your students' experiences in this unit and that invites passersby to interact and post comments and ideas as well. The display should contain samples of students' work that exemplify their strategies, struggles, and questions. With the class, organize the final document in a way that clearly communicates students' development. Here are three big ideas to keep in mind as you organize this material:

- What were the questions students explored? These should be clearly delineated.
- What did students learn? Find pieces of work that clearly represent the ideas developed by the classroom community.
- What questions were raised? Some of these questions have probably been answered; others may still need to be explored. This should be indicated on the document.

Assessment Tips

After the wall has been created, ask students to reflect in writing on their learning over the past two weeks. Ask them to trace their thinking by choosing an idea they introduced. Invite them to examine how and why this idea changed. This written reflection can then be placed in their math portfolios.

Reflections on the Unit

The mathematician Samuel Karlin once wrote, "The purpose of models is not to fit the data but to sharpen the questions." In this unit, the open array was developed as a model for multiplication. This model was developed through a number of different problems, all connected to the context of packing truffles. In thinking about packing truffles in boxes of ten, students explored the multiplicative structure of our base-ten number system. Afterward, students thought about creating box blueprints for different quantities of truffles. The experience of building blueprints laid the foundation for students to think about arrays as tools to model the distributive, associative, and commutative properties of multiplication. As the unit progressed, the open array emerged as a model to represent student strategies in multiplication.

Although this model is situation-specific in this unit, it will, over time and with use, become a generalized model for students to think with. To support this development, use the open array throughout the year as you continue to work on multiplication and division. A resource unit in the *Contexts for Learning Mathematics* series, *Minilessons for Extending Multiplication and Division,* can provide a helpful variety of activities for working with these operations. As you continue to do minilessons with strings of related problems, use the open array to record students' strategies for finding solutions.

MUFFLES' TRUFFLES
218 Raspberry
83 Vanilla cinnamon
46 Almond & raisin

Names ______________________________ Date ____________________

Here are the truffles that Muffles' assistant Patricio needs to package:

- 218 raspberry truffles
- 132 strawberry truffles
- 174 dark chocolate truffles
- 83 vanilla truffles with cinnamon and nutmeg
- 126 green truffles with pistachios
- 308 truffles with pecans and caramel
- 97 butterscotch crunch truffles covered in milk chocolate
- 22 truffles with white and dark chocolate swirls
- 44 chocolate-covered cherry truffles
- 46 almond and raisin truffles

Questions to think about:

- How many boxes does Patricio need for each flavor? How many leftovers of each kind will there be?
- Is there a shortcut way to know how many boxes of each kind he needs to pack and how many leftovers there will be for the assortment boxes?
- How many assortment boxes can he make?
- Muffles sells his fancy truffles for $1.00 each so his boxes of truffles cost $10 each. How much money will he collect if he sells them all?

Appendix C

Each of the problems used in the minilessons on Days Five and Six can be created from the templates below. Make multiple copies on transparency paper and cut them out.

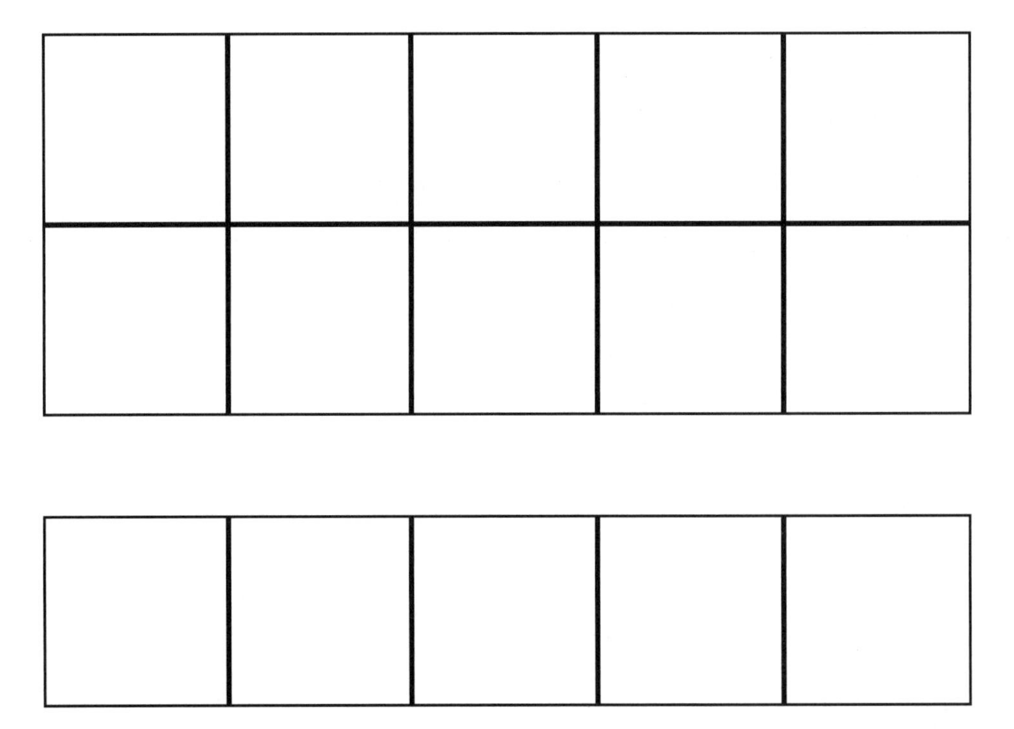

MUFFLES' TRUFFLES
Raspberry
Strawberry
Dark Chocolate
Vanilla Cinnamon
Pistachio
Pecan & Caramel
Butterscotch
White & Dark Chocolate
Chocolate Cherry
Almond & Raisin
MUFFLES' TRUFFLES
$10
$10
$10

Appendix E

2 x 5 box blueprints

Student recording sheet for the new boxes investigation

Names ______________________________ Date ______________

Number of 2×5 Arrays Used	Total Number of Squares	New Blueprints Built
2	20	2×10 4×5

MUFFLES' TRUFFLES

Appendix H

Box outlines

Directions for making box outlines: Using graph paper with one-inch squares, cut out the following arrays:

10 × 10	12 × 12
10 × 12	9 × 6
8 × 7	7 × 7
9 × 7	8 × 8
6 × 7	9 × 9
5 × 6	4 × 4
6 × 6	9 × 8

Now, using these as patterns, trace around each array on blank paper and then make copies of the complete set. Make one complete set of the fourteen box outlines for each pair of students in your class.